T0231923

SOIL PROPERTIES AND THEIR CORRELATIONS

SOIL PROPERTIES AND THEIR CORRELATIONS

SECOND EDITION

Michael Carter
Geotechnical Consultant (Retired), UK

Stephen P. Bentley
Cardiff University, UK

This edition first published 2016
© 2016 John Wiley & Sons, Ltd

First Edition published in 1991

Registered Office
John Wiley & Sons, Ltd, The Atrium, Southern Gate, Chichester, West Sussex, PO19 8SQ, United Kingdom

For details of our global editorial offices, for customer services and for information about how to apply for permission to reuse the copyright material in this book please see our website at www.wiley.com.

Library of Congress Cataloging-in-Publication Data

Names: Carter, Michael, author. | Bentley, Stephen P., author.
Title: Soil properties and their correlations / Michael Carter and Stephen P. Bentley, Cardiff University, UK.
Description: Second edition. | Chichester, West Sussex, United Kingdom : John Wiley & Sons, Inc., 2016. | Includes bibliographical references and index.
Identifiers: LCCN 2016013755 (print) | LCCN 2016014304 (ebook) | ISBN 9781119130871 (cloth) | ISBN 9781119130901 (pdf) | ISBN 9781119130895 (epub)
Subjects: LCSH: Soil mechanics.
Classification: LCC TA710 .B4285 2016 (print) | LCC TA710 (ebook) | DDC 624.1/5136–dc23
LC record available at http://lccn.loc.gov/2016013755

A catalogue record for this book is available from the British Library.

Set in 10/12pt Times by SPi Global, Pondicherry, India

1 2016

Contents

Preface

The aims of this book are to provide a summary and discussion of commonly used soil engineering properties and to give correlations of various engineering properties.

The book includes:

- a compendium of published correlations;
- discussions of the reliability, accuracy and usefulness of the various correlations;
- practical advice on how soil properties are used in the assessment and design of geotechnical problems, including basic concepts, and limitations on their use that need to be considered; and
- descriptions of the measurement of soil properties, and how results are affected by the method of measurement and the expertise of technicians carrying out the testing.

A consideration in describing the various properties has been an awareness by the authors that many geotechnical engineers and engineering geologists have little, if any, hands-on experience of laboratory testing, and are often unaware of the procedures used to obtain the various soil properties and of the effects of poor or inappropriate practice.

The properties are also described in relation to their use in geotechnical analysis, in a way that we hope will give students and younger engineers an in-depth appreciation of the appropriate use of each property and the pitfalls to avoid, and should also provide a useful reminder to more experienced professionals.

Many soil correlations were established in the early decades of soil mechanics, with there being no need to repeat the work once correlations had been established and verified by sufficient researchers. As a consequence, the correlations given in this book span a wide range of time, a few as far back as the 1930s, but we have also presented more recent work where this adds useful information. However, our intention in selecting correlations is to present those that will be of wide practical application, and the book is not intended as a research review. To aid their use in spreadsheet calculations, we have derived mathematical expressions to fit many of the correlations that were originally given only graphically. We have also tried to keep the work independent of national design codes, but it inevitably contains references to practices that are more prevalent in the English-speaking world. Where references are made to classification systems and associated codes we have, where possible, included references to both UK and US practice.

We envisage and recommend that correlations be used in two ways: firstly, to obtain values of a property that has not been measured; and secondly, to provide additional values where some direct measurements of the property have been made. In the first case, where no values of a particular property have been directly measured, the values obtained from correlations should be viewed with caution and treated as preliminary, especially where the property value is critical to the predicted performance of a design. Where correlations are used in combination with direct measurements to provide supplementary values, the accuracy and reliability of the correlations can usually be verified, fine-tuning the correlation if necessary, which may allow the values obtained by correlation to be viewed with more confidence.

While every care has been taken in the preparation of this book, with the very large amount of information that has been assembled it is possible that some errors have occurred; users should satisfy themselves that the information presented is correct. The authors can take no responsibility for consequences resulting from any errors in the book. The views expressed about the reliability and accuracy of correlations, typical values and other published information are based on the authors' own experience and may not accord with those of other geotechnical specialists.

Acknowledgements

In creating a compendium of published correlations, we had to seek permission from many authors around the globe; for her role in this important and painstaking task, the authors would like to thank Carol Clark. Bringing together such a large number of disparate items of information from many sources also involved a great deal of checking, and our thanks go to ex-colleagues Jason Williams and Max Lundie for their checking of some of the work, and especially to Mark Campbell who read through the entire script, noting errors and giving many helpful suggestions.

List of Symbols

Symbol	Name of variable	Typical units (SI)*
α	A scaling factor for estimating footing settlements from plate bearing test results.	D
α	A factor for estimating values of coefficient of volume compressibility from static cone test results.	D
α	Adhesion factor, for pile calculations.	D
α	A factor used to estimate the pull-out resistance of a soil reinforcement grid.	D
Δp	A distance above or below the A-line on a standard plasticity chart.	%
θ	Angle of a plane, from the direction of maximum principle stress, on which stresses act.	Degrees
μ	Viscosity of permeant for general seepage calculations.	kN.s/m^2
ν	Poisson's ratio.	D
π	Ratio of the circumference of a circle to its diameter (≈ 3.14159).	D
ρ	Settlement.	m, mm
σ	Direct stress.	kPa (kN/m^2)
σ'	Effective direct stress.	kPa (kN/m^2)

$\sigma_1, \sigma_2, \sigma_3$	Maximum, intermediate and minimum principal stresses.	kPa (kN/m²)
σ_n	Effective earth pressure, used in soil nail calculations.	kPa (kN/m²)
σ_v, σ'_v	Vertical stress, or overburden pressure, in total and effective stress terms, respectively.	kPa (kN/m²)
τ	Shear stress.	kPa (kN/m²)
γ	Bulk density of soil.	kN/m³
γ_d	Dry density of soil.	kN/m³
γ_{dmax}	Maximum dry density, for relative density calculations.	kN/m³
γ_{dmin}	Minimum dry density, for relative density calculations.	kN/m³
γ_p	Density of permeant for general seepage equation.	kN/m⁴
γ_{sub}	Submerged density of soil.	kN/m³
γ_w	Density of water.	kN/m³
φ	Angle of shearing resistance (general, or in total stress terms).	Degrees
φ'	Effective stress angle of shearing resistance.	Degrees
φ_d	Drained angle of shearing resistance.	Degrees
φ_r	Residual angle of shearing resistance (general).	Degrees
a	Air voids content of soil.	%
a	Component of influence factor I_c for estimating settlements of footings on sands.	D
A	Area (nominal) of soil water flow.	m²
A	A correction factor for rod energy ratio in the standard penetration test.	D
A	Percentage passing a 2.4 mm sieve, used in the calculation of suitability index.	%
A	A constant used in the estimation of swelling potential from plasticity index.	D
A_c	Activity value (of a clay).	D
A_p	End area of penetration cone in a 1standard penetration test.	mm²
A_s	End area of penetration cone in a dynamic probe.	mm²

a_v	Coefficient of compressibility. (See also m_v, coefficient of volume compressibility.)	m^2/MN
b	Component of influence factor I_c for estimating settlements of footings on sands.	D
B	Footing width.	m
B	A constant used in the estimation of swelling potential from plasticity index.	D
c	Shape factor in general seepage calculations.	D
c	Cohesion.	kPa (kN/m^2)
C	Percentage finer than 0.002 mm, used in the calculation of activity for a clay.	D
c'	Effective stress cohesion.	kPa (kN/m^2)
C_1	Constant used in Hazen's formula to estimate the coefficient of permeability.	D
CBR	California Bearing Ratio.	%
C_c	Coefficient of curvature (coefficient of grading).	D
C_c, C_r	Compression index, recompression index, respectively.	D
c_d	Drained cohesion.	kPa (kN/m^2)
CI	Consistency index.	%
C_N	Correction factor for overburden pressure, applied to SPT N-values.	D
c_u	Undrained cohesion, shear strength.	kPa (kN/m^2)
C_u	Coefficient of uniformity.	D
c_v	Coefficient of consolidation.	cm^2/s, $m^2/year$
C_a	Secondary compression index.	$(\log_{10} time)^{-1}$
$C_{a\varepsilon}, C'$	Modified secondary compression index (sometimes referred to simply as the secondary compression index).	$(\log_{10} time)^{-1}$
d	Maximum length of drainage path in consolidation calculations.	m
D	Depth of foundation (when calculating allowable bearing pressures on sands).	m
D_{10}	The 10% particle size, also called the effective size.	mm (or µm)
D_{30}, D_{60}	The 30% and 60% particle sizes, respectively.	mm (or µm)

D_n	The particle size at which $n\%$ of the material is finer. See also D_{10}, D_{30}, D_{60}.	mm (or µm)
D_r	Relative density (of granular soils).	D
D_s	An effective particle size for permeability estimates, usually taken as D_{10}.	mm
e	Voids ratio.	D
e	The natural number, approximately 2.718.	D
E	Young's modulus (also called the elastic modulus).	kPa, MPa
e_1, e_2	Initial and final voids ratios in consolidation testing.	D
E_d	Deformation modulus (also called the constrained modulus).	kPa, MPa
e_{max}	Maximum voids ratio, for relative density calculations.	D
e_{min}	Minimum voids ratio, for relative density calculations.	D
ER_r	Rod energy ratio in standard penetration test.	D
F	The percentage passing the 75 µm sieve, used in the calculation of AASHTO classification group index.	%
f_p, f_s, f_t	Shape, layer thicknes and time factors, respectively, for estimating settlements of footings on sands.	D
F_p	Drop distance of monkey (falling hammer) in a dynamic probe.	mm
F_s	Drop distance of monkey (falling hammer) in a standard penetration test.	mm
G	Shear modulus.	kPa, MPa
G_s	Specific gravity of soil solids	D
h	Thickness of specimen in consolidation testing.	mm
H	Thickness of a compressible layer in consolidation testing.	m
i	Hydraulic gradient in soil water flow.	D
I_c	Influence factor for estimation of settlements of footings on sands.	D

I_r	Rigidity index, used in rate-of-settlement estimates based on static piezocone test results.	D
I_r	Swell index, used in the estimation of swelling pressure.	D
k	Coefficient of permeability.	m/s, m/year
K	A constant used in the estimation of swelling potential from plasticity index.	D
K_0	Coefficient of earth pressure at rest.	D
K_d	Depth factor for allowable bearing pressures on sands.	D
K_s	Earth pressure coefficient use in driven pile calculations.	D
L	Footing length.	m
LI	Liquidity index.	%
LL	Liquid limit.	%
m	Moisture (water) content of soil.	%
M_p	Mass of monkey (falling hammer) in a dynamic probe.	kg
M_s	Mass of monkey (falling hammer) in a standard penetration test.	kg
m_v	Coefficient of volume compressibility. (See also a_v, coefficient of compressibility.)	m²/MN
n	Porosity of soil.	D
n	A factor used to estimate undrained shear strength from consistency index or liquidity index.	D
N	SPT N-value; blows of standard hammer to drive the SPT sampler or cone 300 mm.	Blows
N_1	SPT N-value corrected for overburden pressure.	Blows
$N_{1(60)}$	SPT N-value corrected for overburden pressure and to a rod energy ratio of 60%.	Blows
N_{60}	SPT N-value corrected for rod energy ratio, ER_r. (the "60" refers to standardisation to 60% rod energy.)	Blows
$N_{corrected}$	SPT N-value corrected for silts and fine sands below the groundwater table.	Blows

N_k	A factor used in the estimation of undrained shear strength from static cone tip resistance.	D
O_{40}, O_{80}	Pore diameters at which 40% and 80% of the pores are finer	mm, µm
OCR	Overconsolidation ratio.	D
p	Previous maximum overburden pressure, used in estimating settlements of footings on sands.	D
p_1, p_2	Initial and final pressures used in a stage of consolidation testing.	kPa (kN/m^2)
PI	Plasticity index.	%
PL	Plastic limit.	%
PM	Plasticity modulus.	%
P_p	Penetration for each blow count in a dynamic probe.	mm
P_s	Penetration for each blow count in a standard penetration test.	mm
q	Quantity of flow of water through soil per unit time.	m^3/s, m^3/year
q	Bearing pressure.	kPa (kN/m^2)
q_a	Allowable bearing pressure.	MPa (MN/m^2)
q_c	Measured cone resistance (pressure) in static cone tests.	kPa (kN/m^2)
q_u	Ultimate bearing capacity.	kPa (kN/m^2)
R	Component of influence factor f_t for estimating settlements of footings on sands.	D
S	Degree of saturation.	%
S	Swelling potential.	%
s, s_u	Undrained shear strength.	kPa (kN/m^2)
S_t	Sensitivity	D
SL	Shrinkage limit.	%
t	Time, used in calculations or rates of consolidation and secondary compression.	s, years
t_1, t_2	Start and end times for secondary compression calculations.	s, years
T_v	Basic time factor, used in calculations or rates of consolidation.	D

u	Pore water pressure.	kPa (kN/m^2)
U	Degree of consolidation.	D
v	Nominal velocity of flow of water through soil.	m/s, m/year
v_t	True velocity of flow of water through soil.	m/s, m/year
W_{LW}	Weighted liquid limit, used in the estimation of swelling potential.	%
W_w	Weight of water (in the model soil sample).	g
Y	Rate of frost heave.	mm/day

*D = dimensionless; % values are also essentially dimensionless.

List of Property Values and Correlations in the Tables and Figures

Table	Index properties			Density			Compressibility			Strength		Specialised					Probe testing		
	Grading	Moisture content	Plasticity/ consistency limits	Density	Relative density	Permeability	Total settlement (m_v, C_c and settlement of sands)	Rate of settlement (c_v)	Coefficients of secondary compression	Shear strength	California Bearing Ratio	Shrinkage and swelling characteristics	Frost susceptibility	Susceptibility to combustion	Stresses at soil-structure interfaces	Soil classification	Standard penetration test	Dynamic cone tests	Static cone tests
3.1				✓															
3.2				✓												✓			
3.3				✓												✓			
3.4					✓												✓		
3.5																	✓		
3.6																	✓		
3.7																	✓		
4.1						✓													
5.1							✓									✓			
5.2							✓												
5.3								✓								✓			
5.4							✓		✓							✓			
5.5							✓									✓			✓
6.1										✓						✓			
6.2			✓							✓									
6.3										✓						✓			
6.4			✓							✓						✓			
6.5										✓						✓			
7.1											✓								
7.2			✓								✓					✓			
8.1												✓							
8.2			✓																
8.3												✓							
8.4		✓										✓							
8.5		✓										✓							
8.6	✓	✓										✓							
8.7	✓	✓										✓							
9.1	✓	✓											✓						
9.2													✓			✓			
11.1															✓	✓			
11.2															✓	✓			
A1																✓			
A2																✓			
A3																✓			
A4																✓			
A5																✓			
A6																✓			
A7																✓			
A8																✓			
B1																			✓

Figure	Index properties			Density			Compressibility			Strength		Specialised						Probe testing		
	Grading	Moisture content	Plasticity/consistency limits	Density	Relative density	Permeability	Total settlement (m_v, C_c and settlement of sands)	Rate of settlement (c_v)	Coefficients of secondary compression	Shear strength	California Bearing Ratio	Shrinkage and swelling characteristics	Frost susceptibility	Susceptibility to combustion	Stresses at soil-structure interfaces	Soil classification	Standard penetration test	Dynamic cone tests	Static cone tests
1.13										✓									
2.1	✓															✓			
2.2	✓																		
2.4			✓																
2.5			✓													✓			
3.1				✓															
3.2		✓		✓															
3.4		✓	✓	✓															
3.5		✓		✓															
3.6																	✓		
3.7					✓														✓
3.8																	✓		✓
4.1						✓										✓			
4.2	✓					✓													
5.3			✓				✓										✓		
5.4							✓												
5.5							✓												
5.6								✓											
5.7			✓					✓											
5.9		✓							✓										
5.10		✓							✓										
5.11							✓										✓		
5.12							✓										✓		
5.13							✓										✓		
5.14							✓										✓		
5.15							✓	✓											✓
6.4			✓							✓									
6.5			✓							✓									
6.6			✓							✓									
6.7			✓							✓									
6.8			✓							✓									
6.9										✓							✓		
6.10		✓								✓							✓		
6.11		✓								✓									
6.12				✓	✓					✓							✓		
6.13					✓					✓							✓		
6.14					✓					✓									
7.1		✓	✓								✓								
7.2		✓									✓								
7.3	✓		✓								✓								

Figure	Index properties			Density			Compressibility			Strength		Specialised				Soil classification	Probe testing		
	Grading	Moisture content	Plasticity/consistency limits	Density	Relative density	Permeability	Total settlement (m_v, C_c and settlement of sands)	Rate of settlement (c_v)	Coefficients of secondary compression	Shear strength	California Bearing Ratio	Shrinkage and swelling characteristics	Frost susceptibility	Susceptibility to combustion	Stresses at soil-structure interfaces		Standard penetration test	Dynamic cone tests	Static cone tests
7.4			✓	✓							✓								
7.5		✓		✓							✓								
7.6											✓					✓			
7.7											✓					✓			
7.8											✓								
8.1			✓									✓							
8.2	✓		✓									✓							
8.4	✓		✓									✓							
8.5			✓									✓							
9.3	✓												✓						
9.4	✓												✓						
9.5	✓												✓			✓			
10.1														✓					
11.1										✓					✓				
11.2										✓					✓				
11.3			✓												✓				
11.4			✓												✓				
A1			✓													✓			
A2			✓													✓			
B8																✓			✓

1

Commonly Measured Properties

The purpose of this chapter is to introduce the more commonly measured properties and give outline descriptions of how they are measured. This will allow engineers and geologists who are specifying test schedules, but who may have little or no experience of soils laboratory work, to have a clear understanding of the procedures used to carry out the tests they are scheduling, along with any problems that might occur. This, in turn, should help them to choose the most appropriate tests and to fully appreciate any problems or shortcomings related to the various test methods when appraising the results. It may also give an appreciation of the complexity of some tests to determine seemingly straightforward properties. For clarity, some details have been omitted; test descriptions are not intended to give definitive procedures or to be of sufficient detail to allow them to be used for actual testing. Such details should be obtained directly from the test standards being used, and will normally be the responsibility of the testing laboratory unless specific variations from the standards are required.

Deeper discussions of the nature and meaning of the various properties, and how they relate to other properties, are given at the beginning of subsequent chapters.

Soil Properties and their Correlations, Second Edition. Michael Carter and Stephen P. Bentley.
© 2016 John Wiley & Sons, Ltd. Published 2016 by John Wiley & Sons, Ltd.

1.1 Moisture Content

Moisture content has a profound effect on many properties, and moisture content determinations are carried out as a routine part of many tests; for example, during the determination of shear strength, compressibility, plasticity and California Bearing Ratio (CBR). An especially common use is during density determinations, where it is used to calculate dry density from measurements of bulk density.

To obtain a moisture content value, a soil specimen is simply heated until dry. By weighing the specimen before and after drying, the weight of dry soil and the weight of water driven off can be obtained, and moisture content is obtained from:

$$m = \frac{\text{weight of water}}{\text{weight of dry soil}} \qquad (1.1)$$

expressed as a percentage.

Note that the definition relates moisture content to the weight of soil solids (dry soil) and not to the total weight of the wet sample. This means that for some soils, such as peat, where the weight of water may exceed the weight of soil solids, the moisture content may exceed 100%.

1.1.1   Test Methods

1.1.1.1   Standard Oven Drying

The standard laboratory procedure is by oven drying a specimen of between 30 g (fine-grained soils) and 3 kg (coarse-grained soils) in an open tin or tray at 105-110 °C for 18-24 hours, or until a stable weight is obtained for at least 4 hours. This temperature is high enough to ensure that all free water is driven off but not so high as to break down the mineral particles within the soil. However, some variation in the method may be required for certain soils, especially those containing gypsum or anthracite (coal), which break down chemically at normal oven temperatures. For such soils, lower temperatures are used, typically 60 °C.

1.1.1.2   Quick Methods

Whist the standard oven drying method is satisfactory for normal ground investigation testing, the length of time taken to dry out the specimen can be a problem for quality control of earthworks, where results are needed quickly. To overcome this problem, a number of quick methods have been developed, some of which are outlined below. In all cases, the quick methods should be calibrated against oven-drying values for each soil type as not all methods work with all soils.

• *Microwave oven drying* works well for most soil types provided the soil is microwaved for the appropriate times – see, for example, Carter and Bentley (1986). Ceramic or glass dishes that do not absorb microwaves must be used, and some kind of dummy sample should be included that will continue to absorb microwaves after all the water has been driven off to avoid running the microwave with no load, which can damage it. Note that there is a risk that the dummy sample will get very hot.

• The *'Speedy' moisture tester* consists of a sealed cylindrical pressure flask with a pressure gauge mounted at one end. A fixed weight of soil is put into it along with calcium carbide powder, and the flask is shaken. Reaction of the powder with water in the soil produces acetylene gas, creating a pressure that is proportional to the amount of water in the specimen. The pressure gauge is calibrated directly in percentage moisture content. The tester is quick and simple to operate, requiring no specialised knowledge or equipment, and usually gives reasonably accurate results with granular soils but results can be erratic with clay soils. The method should be calibrated against oven-drying tests for each soil type, and some soils may not give consistent results at all, precluding use of this method.

• *Field density meters*, which measure the transmission or backscatter of radiation through the soil, may also be used to obtain moisture content values. These are an exception to the methods used by the other tests in that the soil is not dried out during testing. The operation of these devices is summarised later in this chapter in the 'Soil density' section.

• *Other methods* include heating the specimen over a hot tray of sand placed on a gas burner, mixing the specimen with methylated spirit (a mixture of methyl and ethyl alcohol) then setting it alight. However, these are rarely used now except in remote field locations where only primitive equipment is available and they are not without risks to the tester if not carried out carefully, so are not described here.

1.2   Grading

Grading, otherwise known as particle size distribution or PSD, gives a measure of the sizes and distribution of sizes of the particles that make up a soil. Grading is arguably the most fundamental of all properties, especially for coarse-grained soils with little or no clay particles.

Particle size distribution is used for a wide variety of assessments, especially where soil is to be used in remoulded form such as fills and embankments, and grading tests are specified in nearly all site investigation test schedules. Uses include: classifying fill materials for design purposes (Appendix A); assessment of permeability and drainage characteristics; and suitability for backfill to pipes. Grading characteristics are more important for coarse-grained soils (sands and gravels); for fine-grained soils (silts and clays), plasticity is more indicative of behaviour but, even for these soils, the proportion of coarser material present is important for assessing properties.

1.2.1 Test Methods

There are two main methods of grading soil.

- *Sieve analysis*: Coarse-grained soils, with soil particles down to 63 μm (fine sand size, defined as below 75 μm in some standards), can be separated out by sieving.
- *Sedimentation analysis*: Below 63 μm, particles are too fine to be sieved, and particle distribution is determined by the rates of settlement of particles suspended in water using Stokes's Law.

1.2.1.1 Sieve Analysis

British Standard (BS) and US sieves are circular with a square mesh; sieve specifications are defined in BS (2014) and ASTM (2015). The larger sizes are usually 300 mm diameter and the smaller sizes 200 mm. They are made to slot together, one on top of the other, to form a 'nest' (see Fig. 1.1) that can be shaken, usually on an electrically powered sieve shaker, but hand shaking may be used. Within a nest, the sieve aperture size decreases from top to bottom. The minimum sieve size is usually 63 μm or 75 μm, depending on the standard used, and a pan is held below the bottom sieve to catch the fines.

For granular soil (sand and gravel), the specimen may be simply sieved dry. However, soil with a significant amount of silt and clay will tend to form lumps which will be retained on the larger size sieves, so the fines must be washed out of the specimen first. This gives rise to two procedures: wet sieving and dry sieving.

Figure 1.1   Nested sieves.

Wet Sieving
The procedure for wet sieving is as follows.

- The soil specimen is oven dried and weighed. A minimum of 200 g, 2 kg or 20 kg is used depending on whether the soil is fine, medium or coarse grained. (The laboratory will decide on this.)
- The dried specimen is sieved through a 20 mm sieve to separate out larger particles. (This step may be omitted for soils not containing particles above 20 mm.) Material retained on the sieve is sieved through larger sieves (>20 mm).
- Material passing the sieve is divided up, usually by riffling, to produce a suitable-sized sample for the smaller sieves.
- After weighing, this sub-sample is placed in a bucket or tray and covered with water containing a dispersing agent (typically sodium hexametaphosphate) and left to stand for an hour.
- The material is then washed, a little at a time, through a 2 mm sieve nested on a 63 µm sieve until the wash water runs virtually clear.
- Now that the silt- and clay-sized particles have been removed from the soil, the two portions (2–20 mm and 63 µm–2 mm) can be oven dried and dry sieved through a nest of appropriate sieve sizes.
- From the total weight of the sample and sub-samples, and the weights on each sieve, the proportions of the various sizes can be calculated.

This procedure appears rather complex for what is essentially a simple sieving process but it is necessary to ensure that sufficient material is used to obtain a representative specimen size for the larger particle sizes while not overloading the smaller sieves, and to wash out any silt- and clay-sized particles as described above. The need to oven dry the specimen initially, and again after washing out the fines, means that the test can take two or three days overall because each oven drying takes typically 12–24 hours with drying taking place overnight.

Dry Sieving
For clean aggregates or soils with minimal fines content, dry sieving may be used, in which the washing procedure is omitted, significantly simplifying the procedure.

1.2.1.2 Sedimentation Analysis

About 15 g of soil fines (<63 μm or <75 μm depending on the test standard) is boiled with distilled water to wet and break up the particles, then hydrogen peroxide is added to remove any organic matter.

The mixture is allowed to stand overnight and is then boiled again to remove the hydrogen peroxide. It is washed with more distilled water and either centrifuged or filtered before being oven dried and weighed.

The sample is shaken or stirred with sodium hexametaphosphate solution for 4 hours to ensure complete dispersion and washed into a 1 litre measuring cylinder. The volume is made up with distilled water to 1 litre and shaken vigorously, with a stopper in the top of the cylinder. As soon as the shaking has ceased and the cylinder stood on a level surface, a stop clock is started.

The rate of sedimentation is measured by one of two methods. The pipette method uses a pipette to take a sample from 100 mm below the surface at set time intervals. This is dried and weighed. The hydrometer method measures the average density of the suspension by inserting a hydrometer (similar to that used for home wine-making) at various time intervals. With either method, the rate of sedimentation is obtained for a range of time intervals. This can be related to the particle size distribution using Stokes's law, which gives a relationship between particle size and its rate of settlement through a liquid. Sedimentation analysis is not strictly comparable with sieving, especially since Stokes's law assumes the particles to be spherical whereas clay particles tend to be plate-like.

Knowing the percentage of clay particles may be useful in some investigations, but normally the important clay properties are better obtained from the liquid and plastic limit test. Given the complexity of the test, its shortcomings and the limited use of the results, it is worth considering whether sedimentation testing is really necessary before ordering it.

1.3 Plasticity

Plasticity is a measure of the range of moisture contents over which a soil is a mouldable solid. It is measured by two tests: the plastic limit test and the liquid limit test. Results are expressed as three moisture content values: the plastic limit; the liquid limit; and the plasticity index, which is the difference between the liquid and plastic limits.

The tests are carried out on only the fine fraction of a soil, which is normally material passing the 425 μm sieve depending on the test standard. Soils with no fines (granular soils) are not mouldable at any moisture content and are simply described as 'non-plastic'.

Plasticity results are not normally used directly in geotechnical analysis, but may be used in conjunction with correlations to infer a wide range of properties. Liquid and plastic limit values are also used in soil classification systems, as described in Appendix A.

1.3.1 Test Methods

1.3.1.1 Liquid Limit

Two methods are available for determining the liquid limit: the traditional Casagrande method and the cone penetrometer method which has largely superseded the Casagrande method in many regions.

In each case, soil must be initially prepared by drying it then sieving it through a 425 μm sieve and discarding the coarser material. About 200 g of soil is normally sufficient. Oven drying is normally used but this can affect results with some soils, especially tropical residual soils and those containing gypsum, for which air drying should be used.

Casagrande Method
The equipment is illustrated in Figure 1.2. Soil is placed in the cup, the surface scraped level and then grooved as illustrated using a special grooving tool. By turning a handle on the device, the cup is repeatedly raised 13 mm and then dropped on to the special rubber base, jarring the soil and causing the groove to close up. The number of blows required to cause the two sides of the groove to just touch over a 13 mm length is recorded. The soil is then

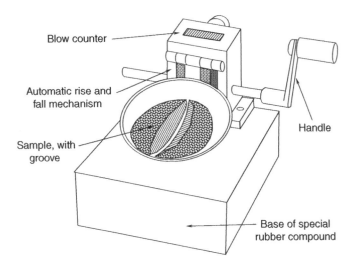

Figure 1.2   Casagrande liquid limit apparatus.

removed and a specimen taken for moisture content determination. The remainder is mixed with a little water and the test repeated. This process is continued until results are obtained for at least four moisture content values, with blow counts of between 50 and 10.

By plotting the number of blows (to a log scale) against moisture content (to a linear scale), the moisture content at which 13 mm of groove would close after 25 blows can be obtained. This is the liquid limit.

Cone Penetrometer Method
The prepared soil is placed in a cylindrical metal cup, scraped level with the rim and placed on the apparatus; the cone is then lowered to just touch the top of the soil, as illustrated in Figure 1.3.

The cone is released for five seconds before being re-clamped, and the cone penetration into the soil is measured to the nearest 0.1 mm with the dial gauge (see Fig. 1.3). The soil is then removed, a specimen taken for moisture content determination, the remainder mixed with a little water and the test repeated. This process is continued until results are obtained for at least four moisture content values, with cone penetrations of between 15 mm and 25 mm.

By plotting the penetration against moisture content (both on linear scales), the moisture content corresponding to 20 mm penetration can be obtained. This is the liquid limit.

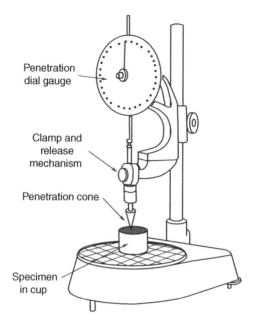

Figure 1.3 Cone penetrometer liquid limit apparatus.

Comparison of the Test Methods

A certain amount of skill and judgement is needed to carry out these tests and to estimate the amounts of water needed. The two tests give similar results but the cone penetrometer method is quicker and produces more consistent results. The Casagrande method, by comparison, can be affected by aging of the rubber base and even by the location of the apparatus on a worktop. For example, if performed on a poorly constructed wooden worktop, tests made with the apparatus mid-way between supports will give slightly different values from those with it placed directly over a support.

1.3.1.2 Plastic Limit

A 20 g specimen of soil, prepared as described for the liquid limit test, is mixed with a little distilled water until it becomes plastic enough to be shaped into a ball. Half the soil is then moulded and rolled between the fingers until the surface begins to crack as it slowly dries out. It is then repeatedly rolled on a glass plate into 3 mm diameter threads until they begin to crack longitudinally at a diameter of 3 mm. The process is repeated with the second half of the specimen and the moisture contents of the two separate portions are determined. The average is taken as the plastic limit provided the two values agree within 0.5%. This moisture content is the plastic limit.

This is an intrinsically simple test, requiring little equipment, but it relies heavily on the experience of the tester. At each re-rolling, the threads must be rolled back into 3 mm threads within 10 hand strokes. The temptation with inexperienced testers is to reduce pressure as the thread begins to break up and continue rolling, giving a plastic limit value that is too low.

1.3.1.3 Plasticity Index

This is simply the difference between the liquid and plastic limits.

1.4 Specific Gravity of Soil Particles

This method is used to determine the average density of the particles of soil. Results are expressed as a proportion of the density of water, that is, as specific gravity.

The specific gravity of the soil particles allows the proportion of voids in the soil to be calculated, and it is needed to calculate the proportion of air voids within a soil, a measure of the effectiveness of compaction methods. However, it is often specified unnecessarily whenever consolidation tests are carried out in the mistaken belief that it is necessary for the calculation of the coefficient of volume compressibility, m_v.

1.4.1 Test Method

The test requires a container that can be filled with water to a high degree of repeatability. Typically, a gas jar, a pyknometer or a specific gravity bottle are used, depending on how coarse-grained the soil is (Fig. 1.4).

The soil specimen is oven dried then weighed. The specimen size varies according to the coarseness of the soil and is typically between 20 g and 400 g. Next, the container is half-filled with water and the soil is placed into it and left to stand for about 4 hours, after which it is shaken vigorously to remove air. It is then filled with water and weighed. Finally, the container is washed out, filled with water and weighed again.

(a)   (b)   (c)

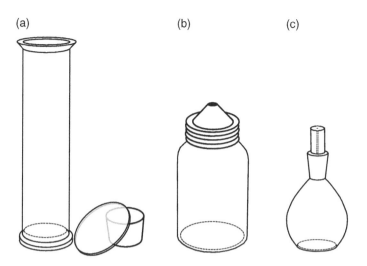

Figure 1.4   Specific gravity determination jars: (a) gas jar with stopper and glass plate; (b) pkynometer; and (c) specific gravity bottle.

The weight of water displaced by the particles can be obtained from the difference between the two container weights and the weight of the dry specimen. The specific gravity G_s is then calculated from:

$$G_s = \frac{\text{weight of soil}}{\text{weight of water displaced by soil particles}}. \qquad (1.2)$$

1.5   Soil Density

Soil density values are used in several ways:

- *directly*, for instance to calculate forces on retaining walls and slopes, and the shear stresses and strengths of soils beneath foundations and behind slopes and retaining walls;
- *indirectly*, for instance to infer other properties such as compressibility and shear strength, using correlations; and
- *for quality control*, for instance to ensure that earthworks and pavements have been compacted as well as reasonably possible.

This wide range of uses means that density measurements are carried out in a variety of situations using various methods.

The presence of water in soil gives an added complication to the definition of soil density, and four types of soil density are used: bulk density, dry density, saturated density and submerged density, as defined and discussed in Chapter 3.

1.5.1   Test Methods

1.5.1.1   Common Laboratory Methods

Density is routinely measured as part of other tests: for instance, shear strength testing using triaxial and shear box equipment; consolidation testing using an oedometer (consolidometer); standard compaction testing; CBR testing; and permeability testing.

In all cases the soil is cut to a specific size or is contained within a mould of fixed dimensions so that determination of the soil volume is a matter of simple calculation. Once the soil is weighed, its bulk density can be readily determined. Once a specimen of the soil has been taken for a moisture content determination, its dry density can also be calculated.

1.5.1.2   Field Density Measurements

These are typically used to check the density of earthworks and pavements for quality control purposes.

Core Cutter Method

This is an intrinsically simple test in which a core of clay is taken by driving a cylindrical core cutter, typically of 150 mm or 200 mm diameter, into the surface. After driving to full depth, it is carefully removed and excess soil is trimmed off either end. Soil density is then obtained from weighings and knowledge of the volume of the cutter.

Care is needed to ensure that the soil completely fills the core cutter, and a special driving dolly is used to help drive the cutter well into the surface while allowing soil to protrude from the top. It also reduces the risk of the sample being compacted during driving which can be a problem with some soils, leading to an overestimation of density.

The method can work reasonably well in firm plastic clays but harder, more friable soils may break up and granular soils simply fall out of the cutter when it is removed. Also, with some soils, the sample may be compacted during driving, leading to an overestimation of density. Because of these difficulties, the range of soils for which this test can be used is limited.

Sand Replacement Method

All the methods described above rely on obtaining a specimen of a specified shape whose dimensions are accurately known. However, this is not always possible when checking the density of compacted earthworks. The sand replacement method overcomes this problem by using sand of known density to fill the volume of an excavated hole.

This, the traditional field test for earthworks, requires no sophisticated equipment; it is however slow and cumbersome, and may appear to be somewhat outdated by modern methods. However, it can be used for many different soil types and gives reasonably accurate values so it is still the standard against which other field density methods are judged. The equipment is illustrated in Figure 1.5.

An area of the surface is scraped level and the special plate with a hole in it (Fig. 1.5) is placed on the levelled surface. A circular hole is excavated using the plate as a template to the full depth of the layer being tested. The plate is used to catch any excavated material which is transferred to a

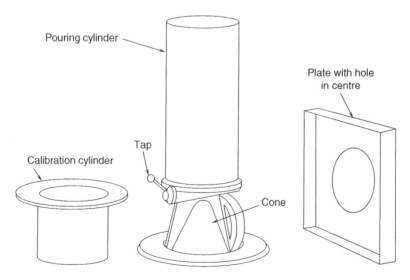

Figure 1.5 Sand replacement method equipment.

polythene bag for weighing and moisture content determination. The hole is trimmed reasonably smooth and any loose material is removed.

The plate is then moved away and the pouring cylinder is placed directly over the hole. This is partly filled with a known weight of special 'density sand' of known density and the tap is opened, allowing sand to flow into the hole, filling both the hole and the cone beneath the cylinder. Once the sand has stopped flowing, the tap is closed and the remaining sand transferred to a polythene bag for weighing.

Knowing the weight of sand used and allowing for the volume of the cone, the volume of soil removed can be calculated, hence the bulk density is obtained. Dry density can then be calculated from the moisture content value.

The density sand, which should be closely-graded (i.e. contain a small range of particle sizes) so that it gives a consistent density value, must be prepared and its density measured beforehand using a special calibration cylinder of known volume.

Equipment comes in various diameters, but is typically 150 mm diameter for fine-grained soils and 200 mm diameter for medium-grained soils. Hole depths are roughly equal to the diameter.

Nuclear Density Meter
The nuclear density meter is quick to use and gives virtually instant results without the need to wait for moisture content determinations. Bulk density is measured using a radioisotope source and a gamma ray detector. The meter is placed on the surface, which must be carefully levelled, and the detector records radiation that has passed through the ground from the source. There are two distinct methods of use, as illustrated in Figure 1.6:

- *direct transmission mode*, in which the source is inserted into the ground and the meter detects radiation that has passed through the soil; and
- *backscatter mode*, in which the source is kept on the surface and the meter detects radiation that has been reflected back.

Backscatter mode is quicker to perform because it does not require a hole to be made in the surface but it effectively checks only the upper few centimetres of soil, whereas direct transmission mode checks the whole layer to the depth of the radioisotope source.

In addition, radiation produces slow neutrons when gamma rays (fast neutrons) collide with hydrogen atoms in the soil. These can be measured

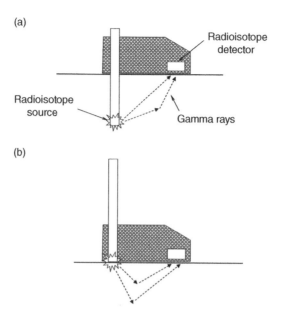

Figure 1.6   Nuclear density meter modes of operation: (a) direct transmission mode; and (b) backscatter mode.

by a slow neutron detector. Since nearly all the hydrogen in soil is normally due to the water present, this can be used to estimate moisture content, allowing dry density to be obtained immediately.

The instruments were originally used to check asphalt densities where the flat surface, lack of need to check a moisture content and requirement for speed make them ideal. However, care must be used in soils, since variations in soil composition can affect results, so they must be checked regularly for each soil type using soil compacted into a calibration box.

There is also a limitation regarding the soil types with which they can be used, especially if the moisture content facility is required, because soils containing organics will contain hydrogen atoms that are not part of the soil moisture. This includes many of the Coal Measures rocks found in Wales and the north of England. Soils containing gypsum will also contain water in the form of water of crystallisation in the gypsum crystals.

A further drawback is that the meters rely on radioactive sources which carry health hazards; although the amount of radioactivity is small, this risk should not be overlooked.

1.6   Permeability

Permeability is used in seepage calculations including seepage through dams, beneath cut-off walls and through the ground when estimating the extent of contamination in the ground. It is therefore limited to certain special types of problem, so permeability tests do not form part of most ground investigation testing.

There are basically two types of laboratory permeability test:

- *falling head tests*, in which a fixed amount of water flows through the specimen from a graduated cylinder, with the pressure head decreasing throughout the test; and
- *constant head tests*, in which water flows under a constant pressure through a soil specimen and the rate of flow is measured.

Falling head tests are more suitable for clay soils, whose low permeability means that the rate of flow is so slow that test times would be excessive. Constant head tests are more suitable for sands and gravels, whose higher permeability means that the falling head test, with its limited supply of water, would be over too quickly for accurate measurement.

In addition to the specific permeability tests described in the following sections, permeability may also be measured using triaxial test equipment and may be inferred from consolidation test results.

Permeability is often measured in field tests where the flow of water from or to a borehole or sometimes a trial pit is measured, again using falling/ rising head or constant head tests depending on the ground permeability. Field testing is beyond the scope of this book, but it is worth considering the relative merits of field and laboratory tests.

Essentially, laboratory tests have the advantage that the soil geometry is controlled so all the parameters needed for permeability calculations are accurately known. This knowledge or control of the ground and the flow through it is absent in field testing, where the ground profile and flow patterns can often only be guessed at. However, the overall permeability of the ground is greatly affected by the macro-structure of the soil, with sometimes thin silt or sand layers within a clay taking most of the flow for instance. The resulting discrepancy can be huge, with overall ground permeability sometimes being 10 times or more that measured in laboratory tests.

These limitations mean that field testing tends to be preferred where the overall permeability of the ground is required, while laboratory testing is preferred to measure the permeability of specific materials such as those to be used as drainage layers.

1.6.1   Test Methods

1.6.1.1   Sample Preparation

Cohesive soil specimens are prepared by carefully trimming a sample to the correct diameter while pressing it into the test cylinder with a cutting edge at one end. Once inserted, the specimen is trimmed top and bottom and parings are taken for moisture content determination. Disturbed cohesive specimens may be formed by first compacting into a standard compaction mould and then forming the test specimen as described above. Granular soils are poured into the permeability test cylinder and tapped or vibrated until the required density is achieved.

Weighings are taken so that the soil density can be determined. Gauze end caps are placed at the top and bottom of the cylinder, which is then fixed into a frame with top and bottom end caps and lowered into a water bath to saturate the soil. A tube connects the upper end of the cylinder to a vacuum pump to aid with soil saturation.

1.6.1.2   Falling Head Permeameter

The specimen, still in the water bath, is connected via tubing to a glass cylinder filled with de-aired water as shown schematically in Figure 1.7.

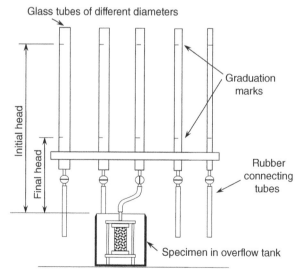

Figure 1.7   Basic layout of the falling head permeameter.

The water is allowed to run through the specimen and the time taken for it to fall between two marks on the graduated cylinder is taken. The soil permeability can then be calculated. The test is usually repeated three or four times.

The apparatus normally has a number of glass tubes of different diameters to allow for variations in permeability of the test specimen. The correct tube is selected by judgement and by trial and error, so that the test will take between about 15 seconds and 90 minutes.

1.6.1.3 Constant Head Permeameter

The sample, still in the water bath, is connected to a column of water maintained at a constant head by the arrangement shown schematically in Figure 1.8. Water flows through the specimen and is collected in a measuring cylinder so that the rate of flow can be determined.

The cylinder containing the soil specimen usually has a number of nipples along its side which can be connected to manometer tubes via flexible hosing, so that the piezometric head can be measured at various points along the specimen.

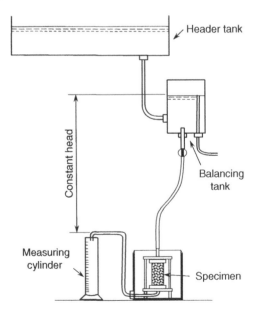

Figure 1.8   Basic layout of the constant head permeameter.

1.7 Consolidation

Consolidation testing is normally carried out by means of an oedometer or consolidometer, shown in Figure 1.9. The test measures two consolidation parameters: the coefficient of volume compressibility m_v, which is a measure of the amount by which a soil will settle; and the coefficient of consolidation c_v, which is a measure of the rate of settlement.

Since settlement of foundations and within fills and embankments is often an important design consideration, consolidation testing commonly forms part of ground investigations.

Experience has shown that values of m_v generally give reasonably good predictions of overall settlement, when appropriate correction factors are applied. It should be remembered, however, that soil becomes stiffer as it compresses, so the value of m_v is not constant but decreases as consolidation pressure increases. It is therefore essential that the m_v value used in settlement estimates is for the pressure change that will actually be experienced in the ground.

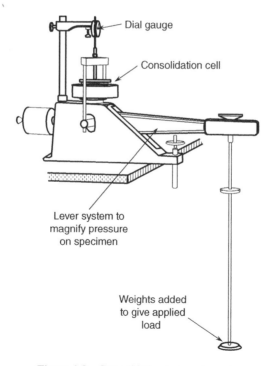

Figure 1.9 Consolidation test equipment.

An exception to this can occur in the case of overconsolidated clays, where m_v will show a sudden increase once the pressure reaches the previous overconsolidation pressure. Care is needed to avoid underestimating settlements in this case.

By contrast, c_v values often do not reflect actual settlement times, especially in highly stratified soils where rates of settlement may be much higher than predicted. This is because the small test specimen does not reflect the macrostructure of the soil, giving the same types of discrepancy as those described for permeability. To overcome this problem and obtain more realistic settlement rates, consolidation tests are sometimes combined with field permeability tests using the theoretical relationships between c_v and the coefficient of permeability.

A further note of caution is needed with some soils and climatic conditions: consolidation theory was developed with temperate conditions in mind, where soils are typically saturated. However, soils such as tropical residual clays and loess tend to be unsaturated, with open structures. This causes specimens to collapse in the initial stages of testing, requiring special testing procedures, and raises questions about the validity of consolidation theory for these soils. Similar problems can occur when testing poorly compacted fills.

1.7.1   Test Method

The soil specimen, usually a clay, is contained in a metal consolidation ring, typically 75 mm diameter and 19 mm deep, with a cutting edge at the bottom. Typically, a short length of sample is extruded from a sample tube and the consolidation ring is pressed into it, cutting around the edges with a palette knife then finally trimming it top and bottom. Good sample preparation is important, and requires care and skill.

For highly stratified soils, where horizontal permeability may be much greater than vertical permeability, pairs of tests may be performed; one with the specimen prepared as described above, and a second specimen in which the consolidation ring is pushed into the side of the sample, so that drainage from it will be along the horizontal layers within the soil. The value of the coefficient of volume compressibility will usually be little affected but the rate of consolidation, hence the coefficient of consolidation, may be greater in the second specimen if the stratification features are fine enough to be reflected in the specimen. However, larger-scale features may not be reflected in the small specimen, and field tests may be used to supplement laboratory testing as described above and in Chapter 4.

Where soil is to be used for fill, a sample can first be prepared by compacting it into a standard compaction mould to the required density then preparing the specimen as described above.

Once prepared, the specimen is placed into a consolidation cell which is fitted on to the load frame as indicated in Figure 1.9. Weights are added, then the consolidation cell immediately filled with water. Readings of displacement with time are taken until the rate of consolidation is very small. More weights are then added and the process repeated; usually four or five load stages are tested, then a final unloading stage may be specified if this is relevant for the design calculations, at which all the weights are removed and swell is measured.

Weights are usually calculated so that the loading pressures follow the sequence: 12.5, 25, 50, 100, 200, 400 kPa (kN/m^2), etc., the choice depending on the expected stress range at the project site. However, it is better to have the first stage below the expected pressure range, as initial settling-in problems can sometimes give less-accurate m_v values for the first load stage.

Readings are typically taken at 30 seconds, then 1, 2, 4, 8, 15 and 30 minutes, then 1, 2, 4, 8 and 24 hours. With five load stages, or four load stages and an unloading stage, the test typically takes a working week.

Graphs are plotted of displacement with time for each load stage, plus a graph of final displacement at each stage against pressure. Graphical constructions are used to obtain c_v.

1.8   Shear Strength

Shear strength testing is usually carried out using either triaxial tests or shear box tests, although other tests, such as ring shear, are occasionally used. Hand penetrometers are also used for quick indications on site, and a variety of shear vane tests may be used including hand-held vanes and vane test equipment for use down boreholes.

For both triaxial and shear box tests, there are a number of variations depending on whether the short- or long-term response of the soil is needed. These are noted in the test methods described below.

Shear strength values are needed for a wide variety of design calculations, including spread footing stability, pile capacity, retaining wall stability and slope stability. Shear strength tests are therefore a feature of most test schedules, although the type of test will depend on both the soils encountered and the design calculations to be carried out.

1.8.1   Test Methods

1.8.1.1   Triaxial Tests

Triaxial tests are usually carried out on undisturbed samples of silt, clay or soft rock. The test is unsuitable for granular soils as the test specimen is not contained within a mould, so non-cohesive specimens cannot be prepared. As noted above, there are a number of variations of the test and triaxial testing is the subject of many books and papers; the complexities of the test procedures and interpretation of results are therefore only discussed briefly in this book.

The apparatus includes a triaxial cell and a load frame, illustrated in Figure 1.10, along with associated control and measuring equipment.

Sample Preparation
Testing is often carried out using samples from 100 mm diameter sample tubes. For fine-grained soils, test specimens are usually 38 mm or 40 mm diameter (76 mm or 80 mm long), and three specimens can be prepared simultaneously by extruding soil from the tube directly into three specimen tubes.

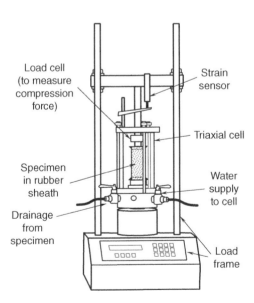

Figure 1.10   Triaxial test equipment.

Specimens are then extruded from the specimen tubes into a special split wall tube to be trimmed to length, then transferred to a third special tube which enables them to be covered in a rubber membrane. An end platen (a disc, which may be either solid or porous depending on the test type) is placed at either end.

In a variation of the test, used for medium-grained soils, a 100 mm diameter specimen is used, straight from the sample tube.

The 'Quick' Undrained Test

This is the simplest, quickest and most common form of the test and is used for footing and pile calculations, for which short-term stability is usually the most critical consideration.

The specimen, with solid end platens, is placed on the pedestal on the load frame and the perspex (plexiglass) triaxial cell is placed over it. Once fixed in place, the triaxial cell is filled with water which is pressurised to simulate pressure in the surrounding ground. The load frame is then adjusted so that the top plunger just touches the top of the specimen. Load is then applied, usually at a constant rate of movement, representing typically 2% strain per minute, up to 20% strain, when the test is normally terminated. Strain and load readings are taken throughout so that a stress-strain plot can be obtained.

The test is repeated on each of the two remaining specimens, each being tested at a different cell pressure. With most commercial laboratory equipment three specimens can be tested simultaneously, each within its own triaxial cell.

Where medium-grained soils are tested and a 100 mm diameter test specimen is used, it is usually not possible to obtain three specimens from a single sample tube. To overcome this problem, a modified form of the test is used in which the specimen is tested up to peak value for the first cell pressure, then the cell pressure is increased and the test continued to obtain a second peak; then similarly for the third cell pressure. This is difficult to perform and the results should be viewed with some scepticism; indeed, some engineers refuse to schedule this type of test.

Drained and Consolidated Undrained Tests

These tests are used to determine the long-term shear strength of the soil after consolidation has taken place. They consequently have to be carried out much more slowly than the 'quick' test, and include an initial consolidation stage after the triaxial cell has been brought up to pressure, before shearing takes place. A further difference from the quick test is that these tests give a measure of the stresses within the soil skeleton (the 'effective' stresses), whereas the quick test measures the combined soil skeleton and pore-water pressure stresses ('total' stresses).

In both tests porous end platens are used so that water can drain out of the specimen during the consolidation phase, which generally takes around 24 hours. In the drained test, drainage of the specimen is also allowed to take place during the loading stage. Loading is carried out sufficiently slowly to allow full dissipation of pore-water pressures during shearing, so the stresses measured represent those on the soil skeleton itself. The rate of testing depends on the permeability of the soil, and laboratories will normally advise on this; shearing typically takes several hours but can, with exceptional soils, take days.

In the consolidated undrained test, the specimen is not allowed to drain during testing so that measured stress includes both the soil skeleton stress and pore-water pressure. However, the pore-water pressure is measured so that, again, the stresses on the soil skeleton can be obtained. The rate of testing needs to be comparable with that of the drained test to allow pore-water pressures to equalise throughout the specimen during shearing, so they can be properly measured.

Specifying Cell Pressures

As described above, each of the three test specimens is tested at a different cell (or confining) pressure and these must be specified when scheduling a test. Pressures should normally cover the range of expected stresses in the soil. The first specimen would therefore be tested at overburden pressure, the third at a little above the maximum expected pressure after loading (e.g. overburden plus additional pressure to be exerted by a foundation), with the second specimen tested midway between. Tests should not be carried out below overburden pressure.

Interpreting the Results

Results are shown as individual stress-strain curves for each specimen and as a graph of shear strength against confining pressure, plotted as Mohr circles, as shown in Figure 1.11. For effective stress tests, additional graphs of the consolidation stage are given. A line is drawn tangential to the Mohr circles, as illustrated, to obtain the shear strength parameters, cohesion c and angle of shearing resistance ϕ (c_d and ϕ_d for drained tests; c' and ϕ' for consolidated undrained tests).

When viewing test results it should be remembered that, while the Mohr circles represent actual test measurements, the tangent line and resulting

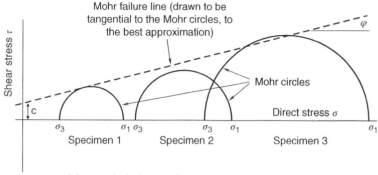

σ_3 minimum principal stress for each test specimen, i.e. the cell pressure
$\sigma_1 - \sigma_3$ pressure applied by the plunger, i.e. the deviator stress
σ_1 maximum principal stress, i.e. cell pressure + deviator stress
c cohesion intercept
φ angle of shearing resistance

Figure 1.11   Example of a Mohr circle plot.

shear strength parameters represent an interpretation; one that may be open to doubt given that variability of the results may make it difficult to decide on where to plot the tangent line. For instance, the plot shown gives both cohesion c and angle of shearing resistance ϕ values but it is typically assumed that for total stresses in saturated clays the angle of shearing resistance is zero, giving a constant shear strength, and that sands and gravels are cohesionless, giving a purely frictional material with no cohesion intercept. Effective strength tests in overconsolidated clays often give both cohesion and friction (i.e. positive c and ϕ values) but the (usually small) cohesion value is sometimes ignored in calculations.

1.8.1.2   Shear Box Tests

The test is straightforward in its approach in that a specimen contained in a square-section split mould (the shear box) is simply sheared across its centre. The specimen and shear box arrangement is shown schematically in Figure 1.12.

Both the test procedure and the theory are simpler than for the triaxial test. Drainage conditions cannot be controlled or measured as much as they can in the triaxial test, and pore-water pressures cannot be measured. However, the test does have two significant advantages over the triaxial test: since the specimen is confined within a mould, it is easy to test granular

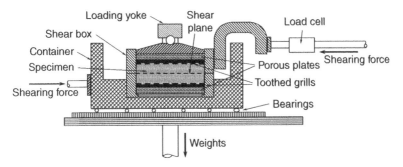

Figure 1.12 Schematic arrangement of a shear box.

soils; and shearing may be continued indefinitely by repeatedly reversing the movement allowing residual strengths to be measured, which is not possible with the triaxial test.

The specimen is usually 60 mm square and 20 mm thick and, for clays, is cut to shape by using a square specimen cutter of the same dimensions as the shear box, trimming with a palette knife. This is then put into the shear box. For sands or disturbed clay samples, the soil may be compacted directly into the shear box.

Once placed in the apparatus, the specimen is put under vertical (or normal) pressure using weights hanging from a loading yoke, and the shear box is flooded with water.

In the standard test, the specimen is sheared at typically 1.25 mm/min, usually for about 9 mm, to obtain peak shear strength. This gives undrained conditions for clays, but drained conditions for sands with their greater permeability. Drained shear strength parameters may be obtained for clays by simply slowing down the rate of shearing, testing over several hours or even days. Residual strength can be obtained by repeated back-and forth shearing. Changes in specimen thickness during shearing (caused by dilation along the shear plane) are measured.

Three test specimens are sheared during a set of tests, each at a different normal pressure, and a plot of shear stress against movement is made for each specimen, with a combined plot of peak (and possibly residual) shear stress against normal stress to obtain the shear strength parameters.

Larger shear boxes, typically 300 mm square, are used to test coarse material, but the large sample size and high normal forces needed create difficulties, especially since the weights required would be too great to be practicable, so a hydraulic loading system is normally used which greatly increases complexity and costs.

1.8.2   Choice of Shear Strength Test

Figure 1.13 gives an overview of the various types of triaxial and shear box test that are commonly used and suggests the appropriate test related to soil types and design considerations. This may provide a useful starting point for the selection of test type but should not be used as a substitute for experience and consideration of specific circumstances.

1.9   Standard Compaction Test

When soil is compacted the density obtained depends on the soil type, its moisture content and the compactive effort used. Standard compaction tests use a standard compactive effort (broadly comparable with that applied by compaction plant on site), and the sample is compacted at various moisture contents to determine the optimum moisture content that produces the maximum dry density. When used for quality control, this gives a standard against which field density results can be judged and an indication of the most appropriate moisture content for compaction. At the site investigation stage, the test gives a comparison between natural moisture content and optimum moisture content, indicating whether the soil is suitable for earthworks; a natural moisture content more than 1½-2% above optimum cannot be well compacted.

1.9.1   Test Method

Soil is compacted into a standard cylindrical mould to give a standard amount of compactive effort per unit volume of soil. A sketch of typical equipment is shown in Figure 1.14. Normally, compaction is carried out using a rammer of standard dimensions, weight and drop, but the use of a rammer for compaction has been found to give unrealistically low values for some sands and gravels; in which case a variation of the test may be used in which the soil is compacted using a vibrating hammer under a specified procedure, typically vibrating for 60 seconds with a force of 30–40 kg.

The standard test uses a 1-litre mould with soil compacted in 3 layers by 27 blows per layer of a 2.5 kg rammer falling through 300 mm; whilst a heavy standard uses a 2.3-litre CBR mould with soil compacted in 5 layers by 62 blows per layer falling through 450 mm. Note that these values are for UK practice, but variations in standards tend to be small and result in much the same compactive effort per unit volume of soil. For instance, the original Proctor test used a $\frac{1}{30}^{th}$ of a cubic foot (944 cm^3) mould with 25 blows per layer of a standard rammer weighing 5½ pounds (2.49 kg) falling through 1 foot (30.48 cm).

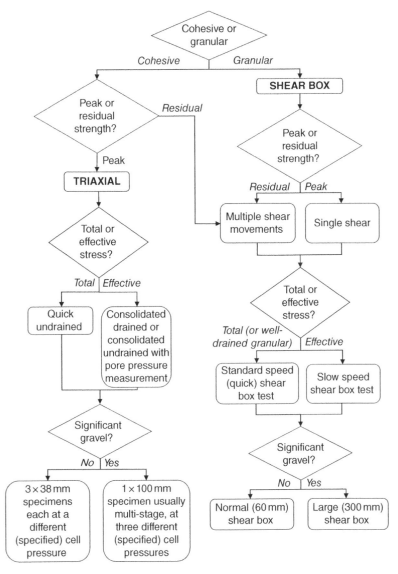

Figure 1.13 Shear test types and typical uses.

Since UK pavement design methods are based on the standard test, there is normally little point in carrying out the heavier standard test in the UK, but design standards elsewhere for both roads and aircraft pavements sometimes require the heavy standard.

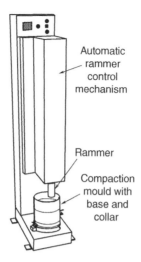

Automatic
rammer
control
mechanism

Rammer

Compaction
mould with
base and
collar

Figure 1.14   Standard compaction test equipment.

About 5 kg of soil is needed for the standard test, and about 12 kg if a CBR mould is used (see Section 1.10 below). The sample of soil is passed through a 20 mm sieve and mixed with a suitable amount of water (or dried out if it is too wet) for the first compaction point. The quantity of water added depends on the soil and must be judged from experience.

The soil is compacted into the mould as described above, using sufficient to fill the mould in three even layers (five for the heavier standard) to just above the base of the collar.

The collar is removed and the top surface carefully trimmed off level with the top of the mould using a steel straight-edge. The mould with contained soil is then weighed. Knowing the weight and volume of the mould allows the bulk density to be calculated. The soil is then extruded from the mould and two or three small specimens are taken for moisture content determination so that the dry density can be calculated.

The remaining soil is mixed with a little water and the test repeated. The test is further repeated, adding water each time, to give at least 5 compaction points. At first, adding water will result in an increase in the dry weight of the soil compacted into the mould, indicating it is dry of optimum. Eventually, however, additional water will result in a decrease in the dry weight compacted into the mould, indicating the soil is wet of optimum. There should be at least two compaction points either side of optimum.

Results are plotted as a graph of dry density against moisture content, on which a trend line may be drawn to obtain the maximum dry density and optimum moisture content. This is discussed further and illustrated in Chapter 3.

Traditionally, earthworks were required to attain a specified percentage of maximum dry density (usually 90% or 95%), but more recently compaction is specified in terms of the percentage of air voids, especially in wet climates where it is impractical to specify that soil be compacted at or near optimum moisture content. This still requires the standard compaction test, however, to determine reasonable, achievable values and compare its moisture content with the optimum value.

1.10   California Bearing Ratio

California Bearing Ratio (CBR) values are used to calculate the required thicknesses of pavement layers. As its name implies, the test was originally developed in California but, ironically, was used there for only a short time. The test does not measure any fundamental property of the soil, and is used only in pavement thickness design.

The test measures the resistance of soil to a cylindrical plunger of standard dimensions being pushed into the surface at a standard rate. It is therefore basically an indirect measure of shear strength, and for clays an approximate relationship exists with shear strength as discussed in Chapter 7.

Results are highly dependent on the method of sample preparation, so much so that the method is no longer used by many highways agencies, but it is still the basis for some design methods, especially for minor roads, car parks and storage areas. The test is intrinsically linked to pavement design methods, and test methods must follow those specified in the design standard being used. For instance, UK design methods generally require the test to be prepared to the moisture content that is likely to exist beneath the pavement in the long term and tested without soaking, whereas US design practice tends to favour soaking of the specimen, regardless of expected subgrade conditions, then applying a climatic correction factor.

1.10.1   Test Method

In the normal laboratory test, a sample is sieved (if necessary) to remove any material above 19 mm size, then compacted into a standard 2.3 litre mould – about 7 kg of sample is normally sufficient. In UK practice, the

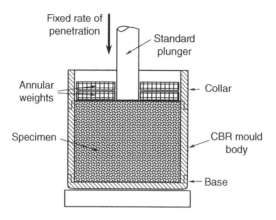

Figure 1.15   Test arrangement for CBR.

target compacted density is usually that which is expected to be achieved in the field, and compaction can be carried out either using a standard rammer (as for the compaction test), a vibrating hammer or by hydraulic jack. Compaction is normally in 3 or 5 layers.

Samples may be soaked, typically for 4 days, during which swell is measured. Soaking is standard in much US practice, but is not normal in the UK.

In the test itself, the plunger is pushed into the surface at 1 mm per minute using a load frame similar to that used for triaxial testing. The arrangement of mould and plunger is shown in Figure 1.15. Readings are taken and a graph plotted of force against penetration. After smoothing and possible corrections, the forces corresponding to penetrations of 2.5 mm and 5.0 mm are compared with a standard value for each penetration and expressed as a percentage; the higher percentage is the CBR value.

During testing (and soaking) annular weights are normally placed on the surface to simulate the weight of pavement that is expected above the layer being tested. These have little effect on clays but may have some effect on granular soil. Typically, two or three 2.27 kg weights are used.

Strictly, in the UK, the sample should be tested at both ends and the average CBR value taken, but some UK laboratories appear to test only at the top. Other design standards may require testing only at the bottom, especially if the specimen is soaked which can cause the top to become softened.

A variation of the test procedure is to use it in conjunction with standard compaction testing: standard compaction tests are performed in CBR moulds rather than standard moulds so that each test specimen can subsequently be tested for CBR. This shows the variation of CBR with moisture content, which can be useful when assessing the suitability of soils for subgrade material.

1.11 Other Properties

A wide variety of other tests are available to check specific properties, and a complete description of all possible tests and test variations is beyond the scope of this book. Three of the more common specialist properties are briefly described below.

1.11.1 Swelling Potential

Swelling potential is used to indicate the potential of a soil to swell and shrink with changes in moisture content. It is used, for instance, in conjunction with estimates of the depth of soil subject to seasonal moisture content changes to determine minimum founding depths for structures on expansive soils and in the proximity of trees.

Swelling potential is defined in terms of the swelling potential test, a modified form of consolidation test, as described in Chapter 8.

An alternative, simpler test, used to assess the propensity of a subgrade soil to swell, is the linear shrinkage test, in which a specimen of soil with moisture content at its liquid limit is pressed into a small trough then oven dried. The reduction in length of the specimen after drying is measured and expressed as a percentage of the original length to give the linear shrinkage value.

Although these tests give an indication of swelling potential, the actual swelling that will occur in the field depends on site conditions including seasonal moisture content changes and overburden pressures, so test results can be used only as an indication of relative swelling problems. Because of this, swelling potential is usually inferred from plasticity and grading tests which are simpler to carry out and based on standard laboratory practice, as discussed in Chapter 8. However, there is no universal agreement about the reliability of plasticity tests to predict swelling potential.

1.11.2 Frost Susceptibility

Frost susceptibility is a measure of the potential of a soil to swell when subjected to repeated cycles of freezing and thawing. It is used, for instance, in conjunction with climate data to determine the minimum thickness of pavements.

Direct measurements of the amount of swelling that can occur are tedious as they require repeated freezing and thawing and, like swelling potential described above, can give only an indication of potential problems, not the specific movement that may occur in the field. Therefore, frost susceptibility is often estimated from plasticity and grading tests, using correlations, as discussed in Chapter 9.

1.11.3 Combustible Content

Combustible content is important when assessing the propensity of a near-surface soil to catch fire when subjected to high temperatures. It is also used in conjunction with air voids content to assess the risk of spontaneous combustion in deeper soils, especially of colliery discard tips.

The propensity for combustion is usually measured by the 'calorific value' test but the simpler 'mass loss on ignition' test may also be used as there is a reasonable correlation between the two tests, as discussed in Chapter 10.

References

ASTM. 2015. Standard specification for woven wire test sieve cloth and test sieves. American Society for Testing and Materials, E11-15.

BS. 2014. Test sieves – technical requirements and testing. British Standards Institution and International Organisation for Standardization, BS ISO 3310-Part 1.

Carter, M. and Bentley, S. P. 1986. Practical guidelines for microwave drying of soils. *Canadian Geotechnical Journal* 23: 598–601.

2

Grading and Plasticity

The concepts of grading and plasticity, and the use of these properties to identify, classify and assess soils, are the oldest and most fundamental in soil mechanics. Their use, in fact, pre-dates the concept of soil mechanics itself: the basic ideas were borrowed from pedologists and soil scientists by the first soils engineers as a basis for their new science.

2.1 Grading

It can be readily appreciated by even the most untrained eye that gravel is a somewhat different material from sand. Likewise, silt and clay are different again. Perhaps not quite so obvious is that it is not just the particle size that is important but also the distribution of sizes that make up a particular soil. Thus, the grading of a soil determines many of its characteristics. Since it is such an obvious property and easy to measure, it is plainly a suitable first choice as the most fundamental property to assess the characteristics of soil, at least for coarse-grained soils. Of course, to rely on grading alone is to overlook the influences of characteristics such as particle shape, mineral composition and degree of compaction. Nevertheless, grading has been found to be a major factor in determining the properties of soils, particularly coarse-grained soils where mineral composition is relatively unimportant.

Soil Properties and their Correlations, Second Edition. Michael Carter and Stephen P. Bentley.
© 2016 John Wiley & Sons, Ltd. Published 2016 by John Wiley & Sons, Ltd.

2.1.1 The Influence of Grading on Soil Properties

During the early development of soil mechanics, engineers relied heavily on past experience and found it convenient to classify soils so that experience gained with a particular type of soil could be used to assess the suitability of similar soils for any specific purpose and to indicate appropriate methods of treatment. Thus, the concept of soil classification arose early in the development of soil mechanics. Even today, despite the development in analytical techniques which has taken place, geotechnical engineers rely heavily on past experience, and soil classification systems are therefore an invaluable aid, particularly where soils are to be used in a remoulded form such as in the construction of embankments and fills.

Poorly graded soils, typically those with a very small range of particle sizes, contain a higher proportion of voids than well-graded soils, in which the finer particles fill the voids between the coarser grains. Grading therefore influences the density of soils (see Table 3.1 in Chapter 3). Another consequence of the greater degree of packing achievable by well-graded soils is that the proportion of voids within the soils is reduced. In addition, although the proportion of voids in fine-grained soils is relatively high, the size of individual voids is extremely small. Since the proportion and size of voids affect the flow of water through a soil, grading can be seen to influence permeability. The theoretical relationship between grading and permeability is discussed in Chapter 4 and the coefficient of permeability is related to grain size in Figure 4.1.

Since consolidation involves the squeezing-out of water from the soil voids as the soil grains pack closer together under load, it follows that the rate at which consolidation takes place is controlled by the soil permeability. Since permeability is, in turn, partly controlled by grading, it can be seen that grading influences the rate of consolidation. Also, since fine-grained soils and poorly graded soils have a higher proportion of voids, and tend to be less well-packed than coarse-grained and well-graded soils, they tend to consolidate more. The consolidation properties of a soil are therefore profoundly influenced by its grading. Since fine-grained soils tend to be more compressible than coarse-grained soils, and consolidate at a much slower rate, it is these soils that are usually of most concern to the engineer in terms of settlement. Their gradings are much too fine to be measured by conventional means and, at these small particle sizes, the properties of the minerals present are of more importance than the grading. Specific correlations between grading and consolidation characteristics do not, therefore, exist. However, the effect of grading on consolidation is taken into account indirectly in some soil classifications which are used to assess the suitability of soils for earthworks and pavement subgrades.

Shear strength is also affected by grading since grading influences the amount of interlock between particles, but correlations between grading and shear strength are not possible because other factors – such as the angularity of the particles, the confining pressure, the compaction and consolidation history, and the types of the clay minerals – are of overriding importance. The variability of some of these factors is reduced where only compacted soils are considered and, with the aid of soil classification systems, the influence of grading on shear strength can be given in a general way as indicated in various tables in Chapter 6. Similarly, the influence of the grading of coarse-grained soils on their California bearing ratio is indicated in Chapter 7 in Table 7.2 and, to some extent, in Figure 7.3.

In a broad sense, both swelling properties and frost susceptibility are influenced by grading. Correlation between grain size and frost susceptibility is described in Chapter 9, but the identification of expansive clays (discussed in Chapter 8) relies almost entirely on the plasticity properties, the only relevant aspect of grading being the proportion of material finer than 425 μm.

2.1.2 Standard Grading Divisions

Although grading, as the most basic of soil properties, is used to both identify and classify soils, the division of soils into categories based on grading varies according to the agency or classification system used. A comparison of some common definitions used is given in Figure 2.1.

This grading-based definition of soil types is usually incorporated into plots of grading tests as indicated in Figure 2.2, which shows soil types against particle sizes for UK practice. The figure illustrates two grading curves:

- curve A: a well-graded soil with a wide range of particle sizes has a moderate slope; and
- curve B: a poorly graded soil, in which one particle size predominates, has a steep slope.

Most natural soils contain a range of soil types – clay, silt, sand, etc. – but the properties of a soil tend to be influenced by the finer particles within it, giving rise to the concept of the effective size, also known as the D_{10} size,

British Standard and MIT

Clay	Silt			Sand			Gravel			Cobb-les	Boulders
	Fine	Medium	Coarse	Fine	Medium	Coarse	Fine	Medium	Coarse		

0.002 0.006 0.02 0.06 0.2 0.6 2 6 20 60 200 mm

Unified soil classification system

Fines (silt and clay)	Sand			Gravel		Cobb-les	Boulders
	Fine	Medium	Coarse	Fine	Coarse		

0.075 0.425 2 4.75 19 75 300 mm

ASTM (D422, D653)

Clay	Silt	Sand			Gravel	Cobb-les	Boulders
		Fine	Medium	Coarse			

0.002 0.075 0.425 2 4.75 75 300 mm

AASHTO (M145)

Silt and clay	Sand		Gravel	Cobbles and boulders
	Fine	Coarse		

0.075 0.425 2 75 mm

Figure 2.1 Some common definitions of soils, classified by particle size.

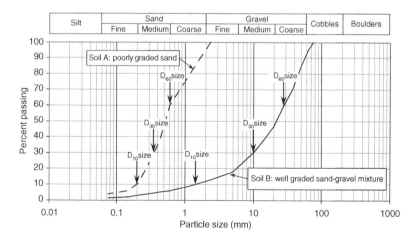

Figure 2.2 Principal features of a grading chart.

which is the particle size for which 10% of the soil particles are finer (see Fig. 2.2). Similar definitions can be used for the particle sizes corresponding to other proportions; for instance, D_{30} and D_{60} are the particle sizes for which 30% and 60%, respectively, of the soil particles are finer, as illustrated in the figure.

In addition, in an attempt to represent how well graded or poorly graded a soil is by a single number, which is useful for specifications, the coefficient of uniformity C_u is often used for predominantly coarser-grained soils, defined as:

$$C_u = \frac{D_{60}}{D_{10}}. \qquad (2.1)$$

Thus, in Figure 2.2: for the well-graded soil of Curve A, $C_u = 28/1.5 = 18.7$ and for the poorly graded soil of curve B, $C_u = 0.6/0.2 = 3.0$.

Several other measures are used in specifications, the most popular of which is probably the coefficient of curvature (sometimes called the coefficient of grading) C_c defined as:

$$C_c = \frac{D_{30}^2}{D_{10} \times D_{60}}. \qquad (2.2)$$

Soils are generally considered to be well-graded if C_u is greater than 6 and C_c is between 1 and 3.

2.2 Plasticity

Just as the concepts of particle size and grading can be readily appreciated for coarse-grained soils, it is obvious that clays are somehow fundamentally different from coarse-grained soils, since clays exhibit the properties of plasticity and cohesion, strictly electrochemical cohesion, whereas sands and gravels do not.

Plasticity is the ability of a material to be moulded without fracturing, while cohesion is the ability to maintain that deformed shape. A good example of these properties is seen when 'throwing a pot' on a potter's wheel. Both properties relate to the electrochemical attraction between the clay mineral particles and are unique to soils containing clay mineral particles. These are plate-like structures, reflecting the molecular structure of the silicate plates from which they are formed. The atomic structure of the plate results in a negative electrical charge on the flat faces of the clay particles and a positive charge at the edges, as illustrated schematically in Figure 2.3a. Electrochemical attraction occurs between the negatively charged clay mineral particles and the positively charged cations, generally either sodium (Na^+) or calcium (Ca^{++}), in the pore water, also indicated in Figure 2.3a. Essentially, a 3-dimensional network of positive-negative attraction binds the particles together, even during deformation.

When the cation concentration in the pore water is low, the negative charge on the clay mineral particles acts over fairly large distances. For example, during formation of the clay, when the particles are sedimented into freshwater, the dominance of the negative face charges causes the clay structure to be semi-orientated or dispersed. A typical clay structure resulting from these conditions is illustrated in Figure 2.3b.

When the cation concentration in the pore water is high, the negative charge on the clay mineral particles is cancelled out over relatively short distances. For example, during formation of the clay when particles are sedimented in sea water, the particles come close together and negative face to positive edge flocculation takes place. A typical clay structure resulting from these conditions is illustrated in Figure 2.3c. This will tend to have a high proportion of voids between the particles but is nevertheless not very porous because the individual voids are very small. Subsequent reductions in cation concentration of the pore water will cause the negative face charge to act over larger distances, which can result in an unstable structure that collapses when disturbed. Examples of such soils are found in sensitive clays, sometimes prone to liquefaction, particularly in Canada and Scandinavia.

Another unique aspect of clay mineral particles is their ability to attract and hold water molecules in 'adsorption complexes' close to the particle surfaces. This creates a thin layer of adsorbed or structured water surrounding the clay particles that has somewhat different properties from free water; for example, it has a lower freezing point. This occurs because water molecules possess a dipole charge, and the positive (hydrogen) side of the molecule is electrochemically attracted to the negative charge on the faces of the clay mineral particles.

Plastic soils (clays) are often described as 'cohesive' to distinguish them from non-plastic soils (sands and gravels), which are described as 'granular' or 'non-cohesive'. The terms 'plastic' and 'cohesive' are often used synonymously. Since all plastic soils are cohesive and all cohesive soils are plastic, this seems quite reasonable. However, electrochemical cohesion and the cohesion that is measured in triaxial tests are fundamentally different. Not all clays that are cohesive necessarily have a measurable cohesion

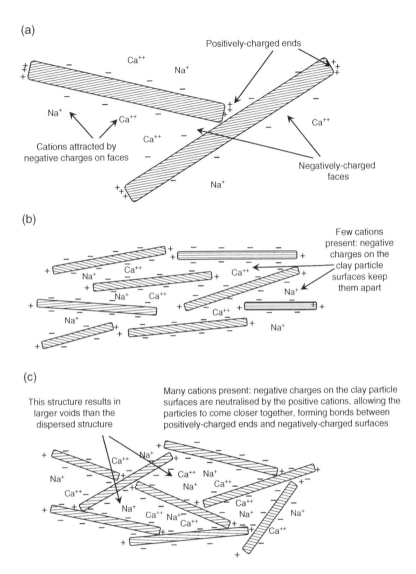

Figure 2.3 Electrochemical bonding between clay mineral particles and the resulting microstructures: (a) electrochemical charges within a clay; (b) semi-orientated (or dispersed) structure; and (c) flocculated structure.

that contributes to their shear strength. Plasticity is a function of the 3-dimensional network of attraction between the negative particles and the positive cations. Electrochemical cohesion is the yield stress that must be applied to overcome the electrochemical forces of attraction and allow movement between the clay mineral particles. The value of this yield stress is insignificant in terms of soil mechanics.

The cohesion that is measured in triaxial tests is generally a function of the negative pore pressures or suction pressures that operate between very-fine-grained particles. The precise mechanism involved is described more thoroughly in Chapter 6.

Interestingly, descriptions of consistency such as 'firm' and 'stiff' were originally defined in terms of plasticity but became redefined in terms of shear strength in the British Standard system. With the change to Eurocodes, definitions of consistencies have reverted back to being based on plasticity.

2.2.1   Consistency Limits

The notion of soil consistency limits stems from the concept that soil can exist in any of four states, depending on its moisture content. This is illustrated in Figure 2.4a, where soil is shown settling out of suspension in water and slowly drying out. Initially, the soil is in the form of a viscous liquid with no shear strength. As its moisture content is reduced, it begins to attain some strength but is still easily moulded; this is the plastic solid phase. Further drying reduces its ability to be moulded so that it tends to crack as moulding occurs; this is the semi-solid phase. Eventually, the soil becomes so dry that it is a brittle solid. Early ideas on the consistency concept and procedures for its measurement were developed by Atterberg, a Swedish chemist and agricultural researcher, in about 1910. In his original work, Atterberg (1911) identified five limits but only three (shrinkage, plastic and liquid limits) have been used in soil mechanics. The liquid and plastic limits represent the moisture contents at the borderline between plastic solid and liquid phases and between semi-solid and plastic solid phases, as indicated in Figure 2.4b. The shrinkage limit represents the moisture content at which further drying of the soil causes no further reduction in volume, as shown in the figure. In electrochemical terms, the clay mineral particles are far enough apart at the liquid limit to reduce the electrochemical attraction to almost zero, and at the plastic limit there is the minimum amount of water present to maintain the flexibility of the bonds.

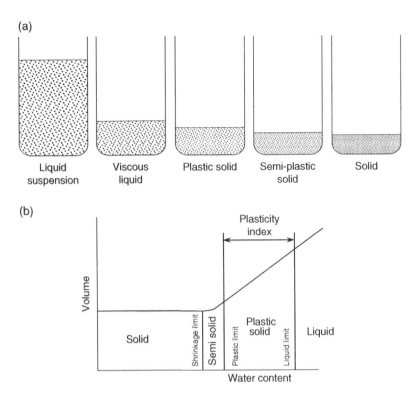

Figure 2.4 Consistency limits: (a) change from liquid to solid as a soil dries out; and (b) volume and consistency changes with water content change.

2.2.2 Development of the Liquid and Plastic Limit Tests

The procedures for liquid and plastic limit determination are described in Chapter 1.

The method of hand-rolling clay into fine threads to determine the plastic limit has remained virtually as it was originally defined in 1910.

The liquid limit test, in which soil was originally held in a cupped hand and tapped gently, evolved to provide much-needed standardisation: a metal dish replaced the cupped hand and the Casagrande apparatus, developed in 1932, replaced the original hand-tapping. Liquid limit determination was further improved when Harrison (1988) suggested a procedure using the cone penetrometer apparatus originally developed for bitumen/asphalt testing.

2.2.3 Plasticity Test Results and Plasticity Descriptions

Liquid limit and plasticity index results are usually plotted on a 'plasticity chart' as illustrated in Figure 2.5. The version shown is used in the UK, but US charts are generally similar (see Appendix A) except that the 'high', 'very high' and 'extremely high' categories are all classified together as 'high', although the precise limits will depend on the standard being used.

As well as illustrating results, the chart can also be used to classify soils into clays (which plot above the A-line) and silts (which plot below the A-line).

2.2.4 The Shrinkage Limit Test

Determination of the shrinkage limit is difficult and results vary according to the test method used and sometimes even depend on the initial moisture content of the test specimen. If the specimen is slowly dried from a water content near the liquid limit, a shrinkage limit value of greater than the plastic limit may be obtained, especially with sandy and silty clays; this is meaningless when considered in the context of Figure 2.3. Likewise, if the soil is in its natural, undisturbed state then the shrinkage limit is often

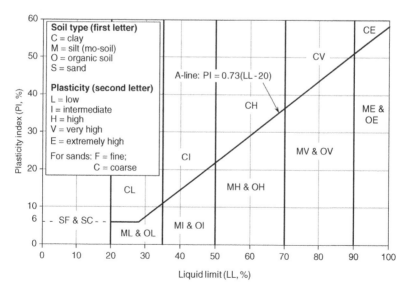

Figure 2.5 Plasticity chart.

greater than the plastic limit due to the soil structure (Holtz and Kovacs 1981; Holtz *et al.* 2011). Karlsson (1977), who carried out shrinkage limit tests on a number of Swedish clays, found that shrinkage limit was related to sensitivity (discussed in Chapter 6). For clays of medium sensitivity the shrinkage limit of undisturbed samples was about equal to the plastic limit, whereas undisturbed highly sensitive clays showed shrinkage limits greater than the plastic limits. Undisturbed organic clays showed shrinkage limits well below the plastic limits. For all the soils tested, the shrinkage limits of the disturbed samples were lower than those of the undisturbed samples, and below the plastic limit.

The sensitivity of shrinkage limit values to the test method is highlighted by Izdebska-Mucha and Wojcik (2013) who compare values obtained using the British Standard method (BS 1990) with those obtained by Polish Standard (PN 1988). For 34 test specimens comprising glacial tills, alluvial soils and Miocene/Pliocene clays, the Polish standard gave shrinkage limits that were an average of 2.6% higher than those obtained using the British Standard, with significant variability; some specimens being over 4% different.

Casagrande suggested that the initial moisture content for shrinkage limit tests should be slightly above the plastic limit, but it is difficult to prepare specimens to such low moisture contents without entrapping air bubbles. It has been found that for soils prepared in this way and that plot near the A-line of a plasticity chart (see Fig. 2.5), the shrinkage limit is about 20. If the soil plots an amount Δp vertically above or below the A-line, the shrinkage limit SL will be:

$$SL = 20 - \Delta p \tag{2.3a}$$

for soils that plot above the A-line and

$$SL = 20 + \Delta p \tag{2.3b}$$

for soils that plot below the A-line.

This procedure to determine the shrinkage limit (for soils prepared in the manner suggested by Casagrande) has been found to be as accurate as the test itself.

An alternative method of estimating shrinkage limit was proposed by Krabbe (Schultze and Muhs 1967) using the relationship

$$SL = PL - 1.25\,PI \tag{2.4}$$

where SL, PL and PI are the shrinkage limit, plastic limit and plasticity index, respectively.

This has been compared against measured values by Izdebska-Mucha and Wojcik (2013), who state that it is a poor predictor of shrinkage limit. However, when comparing predictions with their measured values using British and Polish standards, discussed above, the Krabbe formula seems to give no more variation than that which exists between the two standard test methods.

2.2.5 Consistency Limits as Indicators of Soil Behaviour

The liquid limit should, from the way it is defined in Figure 2.4, be the minimum moisture content at which the shear strength of the soil is zero. However, because of the way the standard liquid limit tests have been defined, the soil actually has a small shear strength. The Casagrande procedure models slope failure due to dynamic loading under quick undrained conditions. The shear strength of the specimen is progressively reduced by increasing its moisture content until a specific energy input, in the form of standard taps, causes a failure of a standard slope in the defined manner. The alternative cone method is also an indirect shear strength test that models bearing failure under quick undrained conditions. The consequence of these test procedures is that all soils at their liquid limit exhibit roughly the same value of undrained shear strength. Casagrande (1932) estimated this value as 2 kPa (kN/m^2), and later work by Skempton and Northey (1952) indicated values of 1–2 kPa. The hand-rolling procedure used in the plastic limit test can be regarded as a measure of the toughness of a soil (the energy required to fracture it) which is also related to shear strength. It has been found that all soils at the plastic limit also exhibit similar values of undrained shear strength, reported by a number of researchers as being 100–200 kPa.

From the preceding discussion it can be seen that all remoulded soils change their strength throughout their plastic range from about 1 kPa at the liquid limit to about 100 kPa at the plastic limit. The plasticity index is therefore the change of water content needed to bring about a strength change of roughly one hundred-fold, within the plastic range of the soil. A remoulded soil with a moisture content within the plastic range can be expected to have a shear strength somewhere between these extremes and it seems reasonable to assume that, for a given soil, its actual shear strength will be related to its moisture content. Also, assuming that the general pattern of shear strength change with moisture content across the plastic range is similar for all soils, then it should be possible to predict the remoulded shear strength of any clay from knowledge of its moisture content and its liquid and plastic limits. Correlations of remoulded shear strength and moisture content, related to the liquid and plastic limit, have been obtained and are discussed in Chapter 6.

A further consequence of these concepts is that a soil with a low plasticity index requires only a small reduction in moisture content to bring about a substantial increase in shear strength. Conversely, a soil with a high plasticity index will not stabilise under load until large moisture content changes have taken place. This implies that highly plastic soils will be less stable and that a correlation may exist between plasticity and compressibility. Also, the liquid limit depends on the amounts and types of clay minerals present, which control the permeability and therefore the rate of consolidation, implying a correlation between liquid limit and the coefficient of consolidation. Correlations between plasticity and consolidation properties are discussed in Chapter 5.

As described above, the special property of plasticity in clays is a function of the electrochemical behaviour of the clay minerals: soils that possess no clay minerals do not exhibit plasticity and, as their moisture content is reduced, they pass directly from the liquid to the semi-solid state. The plasticity limits can give indications of both the type of clay minerals present and the amount. The ratio of the plasticity index to the percentage of material finer than 2 μm gives an indication of the plasticity of the purely clay-sized portion of the soil and is called the 'activity'. Kaolinite has an activity of 0.3–0.5; illite of about 0.9; and montmorillonite of greater than 1.5. These values hold true not only for the activity of the pure clay minerals but also for coarser-grained soils whose clay fraction is composed of these minerals. A high activity is associated with those clay minerals that can absorb large amounts of water within their mineral lattice, and is related to the chemistry of the clay particles. This penetration of the clay minerals by water molecules causes an increase in volume of the clay minerals, so that the soil swells. Activity is therefore a measure of the propensity of a clay to swell in the presence of water and may be used to identify expansive clays. In a less-precise manner, swelling and shrinkage properties are also related to the liquid limit so that this too can be used to help identify expansive clays. This is discussed in Chapter 8.

In broad terms, the plasticity index reflects the ratio of clay mineral to silt and fine sand in a soil; that is, the proportion of clay minerals in the fines. Since the silt-, sand- and clay-sized particles each have their characteristic angles of internal friction, their relative proportions largely determine the angle of internal friction ϕ_r (and therefore, to a large extent, the angle of effective shearing resistance ϕ') of clay soils. There are therefore, perhaps surprisingly, correlations of ϕ_r and ϕ' with plasticity index. These are described in Chapter 6.

2.2.6 Limitations of the Use of Plasticity Limits

It can be seen that, like grading, the plasticity limits are potentially related to a wide variety of soil properties. That this has been found to be true gives ample justification for the use of grading and plasticity properties in the soil classification systems. However, although plasticity limits do give intriguingly good predictions for some engineering properties, certain limitations must be recognised. Limit tests are performed on material finer than 425 µm, and the degree to which this fraction reflects the properties of the soil will depend on the proportion of coarse material present and on the precise grading of the soil.

Another limitation is that the limit tests are performed on remoulded soils and the correlations are not generally valid for undisturbed soils unless the soil properties are not substantially different from the remoulded values. This is the case with many normally consolidated clays, but the properties of overconsolidated clays, sensitive clays and cemented soils often differ markedly from those predicted from the liquid and plastic limit tests.

Further treatment of this subject is given by Haigh et al. (2013).

References

Atterberg, A. 1911. Lerornas forhallande till vatten, deras plastieitetsgranser och plastiitetsgrader (The behaviour of clays with water, their limits of plasticity and their degrees of plasticity). *Kungliga Lantbruksakademiens Handlingar och Tidskrift* 50(2): 132–158; also in *International M itteilungen fur Bodenkunde* I: 10–43.

BS. 1990. *Methods of test for soils for civil engineering purposes*. British Standards Institution, BS 1377.

Casagrande, A. 1932. Research on the Atterberg limits of soils. *Public Roads* 13(8): 121–136.

Haigh, S. K., Vardanega, P. J. and Bolton, M. D. 2013. The plastic limit of clays. *Geotechnique* 63: 435–440.

Harrison, J. A. 1988. Using the BS cone penetrometer for the determination of the plastic limit of soils. *Geotechnique* 38: 433–438.

Holtz, R. D. and Kovacs, W. D. 1981. *An Introduction to Geotechnical Engineering*. Prentice-Hall, New Jersey.

Holtz, R. D., Kovacs, W. D. and Sheahan, T. C. 2011. *An Introduction to Geotechnical Engineering*, 2nd edition. Pearson, New Jersey, USA.

Izdebska-Mucha, D. and Wojcik, E. 2013. Testing shrinkage factors: comparison of methods and correlation with index properties of soils. *Bulletin of Engineering Geology and the Environment* 72: 15–24.

Karlsson, R. 1977. Consistency Limits. Swedish Council for Building Research, Document D6, 40 pp.

PN. 1988. Grunty budowlane Badania probek gruntu. Poliski Komitet Normalizacji Miar I Jakosci. PN-88/B-04481.

Schultze, E. and Muhs, H. 1967. *Bodenuntersuchungen fur Ingenieurbauten.* Springer-Verlag, Berlin.

Skempton, A. W. and Northey, R. D. 1952. The sensitivity of clays. *Geotechnique* 3: 30–53.

3

Density

Density can be defined in two ways:

- *mass density*, mass per unit volume (i.e. kg/m³); and
- *weight density*, often called unit weight or weight per unit volume (i.e. kN/m³).

In practical terms for soil mechanics, where accelerations are not normally applicable, the main difference between the two is simply the units in which density is expressed and, although the term 'unit weight' is often preferred for weight density, the term 'density' is used interchangeably in this text to mean either mass density or weight density.

3.1 Density in the Context of Soils

Since soil is not a homogeneous material but contains three materials in three different phases – soil particles (solid phase), pore water (liquid phase) and air (gas phase) – density is not a single measure of the material but may be defined in essentially two ways:
- *dry density*, which is the weight of soil solids per unit volume, ignoring any water, and

Soil Properties and their Correlations, Second Edition. Michael Carter and Stephen P. Bentley.
© 2016 John Wiley & Sons, Ltd. Published 2016 by John Wiley & Sons, Ltd.

- *bulk density*, which is the total weight of a specimen of soil per unit volume, including any moisture contained within the soil.

It should be noted that quoting a dry density for a soil does not imply that the soil is dry; only that the weight of any water is ignored.

Two further measures of density are:

- *saturated density*, which is the bulk density of a soil in which the soil voids are entirely filled with water, with no air present; and
- *submerged density*, which is the effective density of a soil that is below the water table.

3.1.1 Density Relationships

The relationships between these various densities are best considered by using the concept of the model soil sample, in which a unit volume of solids has been consolidated into one lump, as shown in Figure 3.1. From the model soil sample:

$$\text{dry density, } \gamma_d = \frac{\text{weight of solids}}{\text{volume}} = \frac{G_s \gamma_w}{1+e} \quad \text{and} \tag{3.1}$$

$$\text{bulk density, } \gamma = \frac{\text{wt solids+wt water}}{\text{volume}} = \frac{G_s \gamma_w + \gamma_w Se}{1+e} = \frac{(G_s + Se)\gamma_w}{1+e} \tag{3.2}$$

where notation is defined in Figure 3.1.

The maximum bulk density that a soil can attain for a given voids ratio is the saturated density γ_{sat} when all voids are filled with water. This is obtained by substituting $S=1$ in Equation (3.2).

Also, from the model soil sample shown in Figure 3.1, it can be seen that moisture content can be expressed in two ways: as the volume of water present in the voids multiplied by its density; and as a proportion of the weight of solids. This gives rise to the equivalence, developed in the figure, of:

$$S\ e = mG_s. \tag{3.3}$$

Substituting for this in Equation (3.2) gives:

$$\gamma = \frac{G_s(1+m)\gamma_w}{1+e}. \tag{3.4}$$

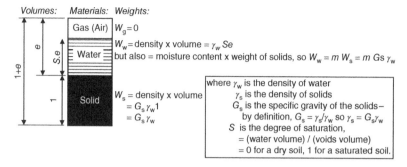

Figure 3.1 Theoretical relationships between density, porosity, voids ratio and moisture content.

This is true for both unsaturated conditions and saturated conditions (when $\gamma = \gamma_{sat}$).

Comparing the bulk and dry densities in the above expressions, we obtain:

$$\frac{\gamma}{\gamma_d} = \frac{G_s(1+m)\gamma_w}{1+e} \times \frac{(1+e)}{G_s \gamma_w} = (1+m), \qquad (3.5)$$

that is,

$$\gamma = \gamma_d (1+m)$$

3.1.1.1 Submerged Density

Since the submerged weight of an object is equal to its weight less the weight of water it displaces then, in density terms:

$$\gamma_{sub} = \gamma - \gamma_w. \qquad (3.6)$$

3.1.1.2 The Effect of Air Voids

Clearly, if a soil contains entrained air then, for a given moisture content, its density will be less than if it were saturated. Referring to Figure 3.1 and defining the air voids content a as the proportion of air voids to the total volume gives:

$$a = \frac{e - Se}{1+e} = \frac{(1+e)-(1+Se)}{1+e} = 1 - \frac{(1+Se)}{(1+e)}.\tag{3.7}$$

Combining Equation (3.7) with Equations (3.1) and (3.3) gives $(1+e) = G_s \dfrac{\gamma_w}{\gamma_d}$, giving

$$a = 1 - \frac{(1+mG_s)}{G_s \dfrac{\gamma_w}{\gamma_d}} = 1 - \frac{\gamma_d(1+mG_s)}{G_s \gamma_w}.\tag{3.8}$$

Rearranging:

$$\gamma_d = \frac{G_s \gamma_w (1-a)}{1+mG_s},\tag{3.9}$$

which allows the dry density to be determined for any given moisture content and proportion of air voids. An example of this is shown in Figure 3.2, where dry density is plotted for zero, 5% and 10% air voids and

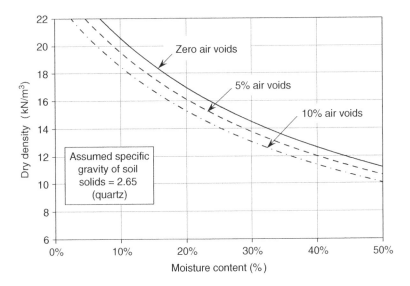

Figure 3.2 Relationship between dry density, moisture content and percentage of air voids.

the zero air voids content. The zero air voids line is often shown on dry density verses moisture content plots used for compaction test results. Other values such as the 5% air voids line may also be useful if, for instance, compacted density is specified in terms of air voids content.

3.1.2   Typical Natural Density Values

Typical values of natural density are given for various soil types in Table 3.1.

3.2   Compacted Density

The compacted density of a soil is not a fundamental property of the soil but depends on the way in which the compaction was carried out. Standard compaction tests are designed to use a similar compactive effort, in terms of compactive energy per unit volume of soil, to that delivered by typical construction plant. The test method, its variations and its applicability to design procedures are discussed in Chapter 1.

Table 3.1   Typical values of natural density.

Material		Natural density	
		Bulk* (kN/m^3)	Dry (kN/m^3)
Sand and gravel:	very loose	17–18	13–14
	loose	18–19	14–15
	medium dense	19–20	15–18
	dense	20–22	17–20
	very dense	22–23	20–22
Poorly graded sands		17–19	13–15
Well-graded sands		18–23	14–22
Well-graded sand–gravel mixtures		19–23	15–22
Clays:	unconsolidated muds	16–17	9–11
	soft, open-structured	17–19	11–14
	firm or stiff normally consolidated	18–22	13–19
	glacial till (boulder clay)	20–24	17–22
Tropical red clays		17–21	13–18
Soil solids	Specific gravity = 2.65 for quartz; 2.64–2.71 for calcareous sand; 2.67–2.73 for most clay minerals.		

*Assumes saturated or nearly saturated conditions.

3.2.1   Typical Compacted Density Values

The compacted density achieved for a soil depends on the soil type, its moisture content and the compactive effort used. Table 3.2 shows typical values
of maximum dry density (MDD) and optimum moisture content (OMC) for
soil classes, using the Unified classification system, for soils compacted to
AASHTO (AASHTO T99; 5.5lb rammer method) or BS (BS 1377: 1990
Test 12; 2.5kg rammer method) standard compaction. The values given are
based on typical values given by Krebs and Walker (1971) and the US Army
Engineer Waterways Experimental Station (Al-Hussaini and Townsend
1975), and on the authors' own experience. A similar set of values but related
to the AASHTO soil classification system is given in Table 3.3. These are

Table 3.2   Typical compacted densities and optimum moisture contents for soil
types using the Unified soil classification system. Adapted from Al-Hussaini
and Townsend (1975).

Soil description	Class	MDD* (kN/m^3)	OMC* (%)
Gravel–sand mixtures:			
well-graded, clean	GW	20–21.5	8–11
poorly-graded, clean	GP	18.5–20	11–14
well-graded, small silt content	GM	19–21.5	8–12
well-graded, small clay content	GC	18.5–20	9–14
Sands and sandy soils:			
well-graded, clean	SW	17.5–21	9–16
poorly-graded, small silt content	SP	16–19	12–21
well-graded, small silt content	SM	17.5–20	11–16
well-graded, small clay content	SC	17–20	11–19
Fine-grained soils of low plasticity:			
silts	ML	15–19	12–24
clays	CL	15–19	12–24
organic silts	OL	13–16	21–32
Fine-grained soils of high plasticity:			
silts	MH	11–15	24–40
clays	CH	13–17	19–36
organic silts	OH	10.5–16	21–45

* MDD and OMC are the maximum dry density and optimum moisture content,
respectively, for the AASHTO or BS standard compaction test AASHTO T99
(5.5lb rammer) or BS Test 12 (2.5kg rammer methods).

Table 3.3 Typical compacted densities and optimum moisture contents for soil types using the AASHTO soil classification system. Adapted from Gregg (1960).

Soil description	Class	MDD* (kN/m³)	OMC* (%)
Well-graded gravel–sand mixtures	A-1	18.5–21.5	5–15
Silty or clayey gravel and sand	A-2	17.5–21.5	9–18
Poorly graded sands	A-3	16–19	5–12
Silty sands and gravels of low plasticity	A-4	15–20	10–20
Elastic silts, diatomaceous or micaceous	A-5	13.5–16	20–35
Plastic clay, sandy clay	A-6	15–19	10–30
Highly plastic or elastic clay	A-7	13–18.5	15–35

*MDD and OMC are the maximum dry density and optimum moisture content, respectively, for the AASHTO or BS standard compaction test (AASHTO T99, 5.5 lb rammer; or BS Test 12, 2.5 kg rammer methods).

based on the above values and the relationship between the AASHTO and Unified soil classification systems, and on values suggested by Gregg (1960).

It should be noted that clean sands often show no clear optimum moisture content and that peak density may be achieved when the sand is completely dry. This is illustrated in Figure 3.3.

Work carried out by Morin and Todor (1977) on red tropical soils in Africa and South America gave correlations between optimum moisture content and plastic limit and between maximum dry density and optimum moisture content, as indicated in Figure 3.4. Morin and Todor also produced a relationship between optimum moisture content and the percentage of particles finer than 2 μm but this showed too wide a scatter to be of use and has therefore not been included in this text.

3.2.2   Quick Estimates of Maximum Dry Density and Optimum Moisture Content

Work carried out by Woods and Litehiser (1938) in Ohio indicated that, for Ohio soils, nearly all moisture-density curves have a characteristic shape. On the basis of over 10,000 tests, 26 typical curves were produced as shown in Figure 3.5. Use of the curves allows the maximum dry density and optimum moisture content to be estimated from a single point on the curve, greatly reducing time and effort for testing compared with the five points

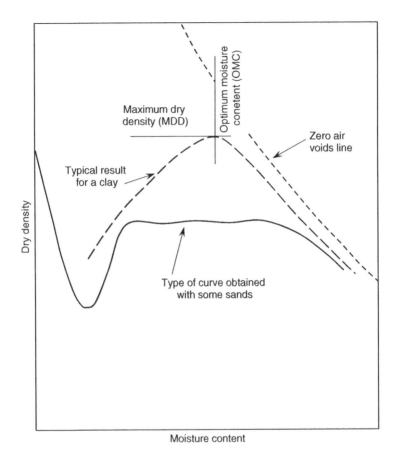

Figure 3.3 Type of density–moisture relationship obtained in the standard compaction test with some sands.

normally tested. It should be noted that the curves are plots of bulk density, instead of the more usual dry density, against moisture content. The inset table gives the corresponding maximum dry density and optimum moisture content for each curve.

These curves can be used to provide quick and fairly accurate estimates of maximum dry density and optimum moisture content, especially when used with rapid moisture content determinations. They have been found to be applicable to soils in many areas, though minor modifications have sometimes been necessary. It is therefore necessary to initially check these curves using full compaction tests and modify as required. Accuracy is

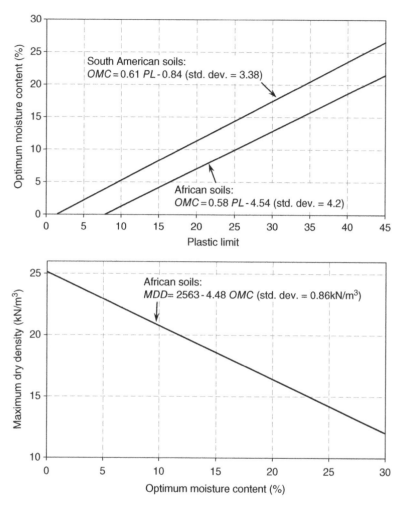

Figure 3.4 Relationships between optimum moisture content, maximum dry density and plastic limit for red tropical soils. Reproduced from Morin and Todor (1977).

improved if the moisture content of the test specimen is close to optimum and preferably on the dry rather than the wet side. The curves are not valid for unusual materials such as uniformly graded sand, highly micaceous soils, diatomaceous earth, volcanic soils or soils in which the specific gravity of the solids differs greatly from 2.67.

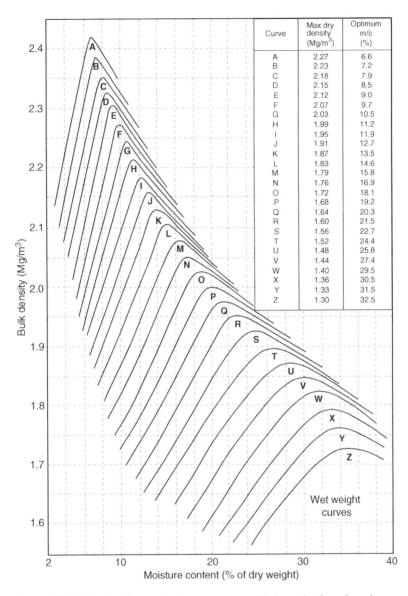

Curve	Max dry density (Mg/m³)	Optimum m/c (%)
A	2.27	6.6
B	2.23	7.2
C	2.18	7.9
D	2.15	8.5
E	2.12	9.0
F	2.07	9.7
G	2.03	10.5
H	1.99	11.2
I	1.95	11.9
J	1.91	12.7
K	1.87	13.5
L	1.83	14.6
M	1.79	15.8
N	1.76	16.9
O	1.72	18.1
P	1.68	19.2
Q	1.64	20.3
R	1.60	21.5
S	1.56	22.7
T	1.52	24.4
U	1.48	25.8
V	1.44	27.4
W	1.40	29.5
X	1.36	30.5
Y	1.33	31.5
Z	1.30	32.5

Figure 3.5 Typical moisture–density curves for rapid determination of maximum dry density and optimum moisture content. Adapted from Woods and Litchiser (1938). Reproduced with permission of Ohio State University.

3.3 Relative Density

For granular soils, relative density is often considered to be a more important indicator of soil properties than actual density. Relative density is defined as:

$$D_r = \frac{e_{max} - e}{e_{max} - e_{min}} \tag{3.10}$$

which, using the relationship given previously between voids ratio and dry density, Equation (3.1), gives, in dry density terms:

$$D_r = \frac{\gamma_{d,max}}{\gamma_d} \times \frac{\gamma_d - \gamma_{d,min}}{\gamma_{d,max} - \gamma_{d,min}} \tag{3.11}$$

where e, e_{min} and e_{max} are the voids ratios in the field and at the densest and loosest states of packing, respectively, and γ_d, $\gamma_{d,max}$ and $\gamma_{d,min}$ are the dry densities in the field and at the densest and loosest states of packing, respectively.

Widely accepted terms for the ranges of relative density D_r are given in Table 3.4.

3.3.1 Field Measurement of Relative Density

Because of the difficulty of obtaining undisturbed samples in sands and gravels, hence of measuring field densities, values are usually estimated from standard penetration test (SPT) results or other probe tests. The classification of relative density based on SPT N-values, shown in Table 3.4, is widely used.

Table 3.4 Relative density descriptions and correlation with SPT N-values in sands and gravels.

Relative density D_r	Description	SPT N-value (blows)
<0.2	Very loose	<4
0.2–0.4	Loose	4–10
0.40–0.6	Medium dense	10–30
0.6–0.8	Dense	30–50
>0.8	Very dense	>50

3.3.2   SPT Correction Factors

3.3.2.1   Correction for Overburden Pressure

Work on sands by Gibbs and Holtz (1957) indicated that the relationship between relative density and SPT N-values (see Appendix B) depends on the characteristics of sand, whether it is dry or saturated and on the over-burden pressure. This led to the suggestion that correction factors (C_N) for overburden pressure should be applied in the determination of relative density and for foundation calculations.

Table 3.5 lists a number of correction factors published by different research-ers, and the results are plotted in Figure 3.6. The corrected value of N is then:

$$N_1 = C_N N. \tag{3.12}$$

It can be seen that the correction factors are in broad agreement except for the Gibbs and Holtz values and, to a lesser extent, the Peck and Barazaa values.

Skempton's factors, given in Table 3.5 are recommended for use in British Standard BS EN ISO 22476-3 (2011), which also suggests the pos-sible use of Liao and Whitman's correction factor.

3.3.2.2   Correction for Fine Sands Below the Water Table

Another correction often applied to SPT values when assessing the relative density of silts and fine sands below the water table is:

$$N_{corrected} = 15 + \frac{1}{2}(N - 15) \tag{3.13}$$

with no correction for N-values of less than 15. This is based on the work of Terzaghi and it is suggested that, because of the low permeability of such soils, pore-water pressures build up during driving of the sampler, resulting in increased N-values. This approach is recommended by Tomlinson (1980) in his discussion of the application of corrections to SPT N-values.

3.3.2.3   Correction for Energy Loss in the System

Although SPT correction factors were discussed at some length by Liao and Whitman (1986), the definitive work on the subject is that of Skempton (1986). Skempton points out that in carrying out the SPT test the energy delivered to the sampler, and therefore the blow count obtained in any given sand deposit at a particular effective overburden pressure, can still vary depending on the method of releasing the hammer, on the type of anvil and

Table 3.5 Summary of published SPT correction factors for overburden pressure.

Reference	Correction factor, C_N		Units of effective overburden pressure σ'_v
Gibbs and Holtz (1957) (equation by Teng 1962)	$C_N = \dfrac{50}{10 + \sigma'_v}$		lb/in^2
Peck and Barazaa (1969)	$C_N = \dfrac{4}{1 + 2\sigma'_v}$ $C_N = \dfrac{4}{3.25 + 0.5\sigma'_v}$	for $\sigma'_v \leq 1.5$ klb/foot^2 for $\sigma'_v > 1.5$ klb/foot^2	klb/foot^2 (1000 lb/foot^2)
Peck *et al.* (1974)	$C_N = 0.77 \log_{10}\left\{\dfrac{20}{\sigma'_v}\right\}$		kg/cm^2 or tons/foot^2
Seed (1976)	$C_N = 1 - 1.25 \log_{10}\left\{\dfrac{20}{\sigma'_v}\right\}$		kg/cm^2 or tons/foot^2
Tokimatsu and Yoshimi (1983)	$C_N = \dfrac{1.7}{0.7 + \sigma'_v}$		kg/cm^2 or tons/foot^2
Liao and Whitman (1986)	$C_N = \sqrt{\dfrac{98}{\sigma'_v}}$		kPa (kN/m^2)
Skempton (1986)	$C_N = \dfrac{200}{100 + \sigma'_v}$ $C_N = \dfrac{300}{200 + \sigma'_v}$ $C_N = \dfrac{170}{70 + \sigma'_v}$	for fine sands of medium D_r for dense, coarse sands, normally consolidated for overconsolidated fine sands	kPa (kN/m^2)

on the length of rods, if less than 10 m. His suggestion is that N-values measured by any particular method should be normalised to some standard rod energy ratio (ER_r), and a value of 60% is proposed. A summary of rod energy ratios for a range of hammers and release methods (with rod lengths >10 m) is given in Table 3.6. N-values measured with a known or estimated ER_r value can be normalised by the conversion:

$$N_{60} = N\left(\frac{ER_r}{60}\right)A \qquad (3.14)$$

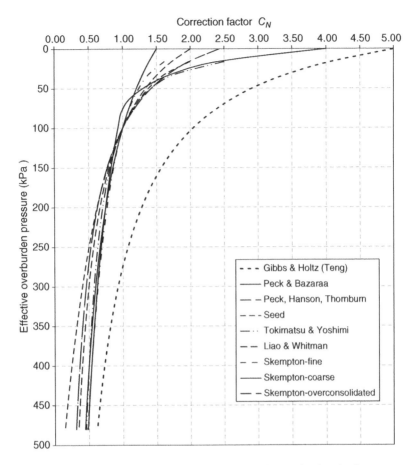

Figure 3.6 Plots of SPT correction factors for overburden depth.

Table 3.6 Summary of rod energy ratios for standard penetration tests. Adapted from Skempton (1986).

Country	Hammer	Release mechanism	ER_r (%)	$ER_r/60$
Japan	Donut	Tombi	78	1.3
	Donut	2 turns of rope	65	1.1
China	Pilcon type	Trip	60	1.0
	Donut	Manual	55	0.9
USA	Safety	2 turns of rope	55	0.9
	Donut	2 turns of rope	45	0.75
UK	Pilcon, Dando	Trip	60	1.0
	Old standard	2 turns of rope	50	0.8

Table 3.7 Approximate corrections A to measured SPT N-values. Adapted from BS (2011).

Influencing factor		Correction factor A
Rod length:	>10 m	1.0
	6–10 m	0.95
	4–6 m	0.85
	3–4 m	0.75
Standard sampler		1.0
US sampler without liners		1.2
Borehole diameter:	65–115 mm	1.0
	150 mm	1.05
	200 mm	1.15

where A represents other correction factors as listed in Table 3.7. The application of the rod length factors given in the table is suggested by British Standard BS EN ISO 22476-3 (2011).

Skempton (1986) states that the Terzaghi and Peck (1967) limits of blow count for various grades of relative density, as enumerated by Gibbs and Holtz, appear to be good average values for normally consolidated natural sand deposits, provided that blow counts are corrected for overburden pressure (N_1) and normalised to a 60% rod energy ratio $(N_1)_{60}$.

3.3.2.4 Appraisal of the Application of Correction Factors

In considering the correction factors it is worth reflecting that the standard penetration test is a fairly crude method of assessing soil properties, especially as it uses the resistance of the ground to dynamic forces (blows) in order to obtain soil parameters such as relative density, shear strength and compressibility that relate to static conditions. In addition, there is an inherent variability in SPT N-values resulting from differences in practice between drillers. This is particularly true for tests carried out in sands below the water table where, in order to obtain meaningful results, the borehole should be kept surcharged with water above the groundwater level at all times. This is often neglected both because it requires a large supply of water and simply out of ignorance. Consequently, groundwater may flow into the borehole, loosening the sand and resulting in artificially low N-values. Alternatively, unrealistically high N-values may be obtained if drillers drive the casing ahead of the borehole to reduce the problem of sand washing up the casing, thus compacting the sand beneath.

Compared with these problems, the corrections relating to the equipment configuration, rod energy losses and borehole size are relatively small. It is therefore unsurprising that these corrections are generally neglected.

The corrections due to overburden pressure and for silts and fine sands below the water table are much greater and are more often used. However, some engineers refuse to use any corrections, arguing that the N-values reflect actual soil conditions including the effects that overburden pressure will have on the soil properties generally.

3.3.3   Other Dynamic Cone Tests

Dynamic probes come in a variety of cone sizes, hammer weights and hammer drops, but modern probes are generally the heavy probe (DPH) and the super-heavy (DPSH) probe. Results are typically expressed as blows per 100 mm.

The super-heavy probe has the same dimensions as an SPT, and interpretation is the same as for SPT data, simply by combining results for 100 mm penetration to obtain the number of blows for each 300 mm penetration. Alternatively, the blows for each 100 mm penetration can be multiplied by 3.

For other size probes, an equivalent SPT N-value may be obtained for granular soils by assuming that penetration is proportional to the energy per blow (hammer mass × fall) and inversely proportional to the cone area. That is:

$$SPT \ N \text{ value} = \frac{M_p F_p A_s P_s}{M_s F_s A_p P_p} \times \text{probe blow} \qquad (3.15)$$

where

M_p and F_p are the mass and fall, respectively, of the probe monkey (the falling weight that drives the probe), A_p is its end area; and P_p is the penetration for each blow count for the probe; and

M_s and F_s are the mass and fall, respectively, of the SPT monkey, A_s is its end area; and P_s is the penetration for each blow count for the SPT.

For an SPT, $M_s = 65 \, \text{kg}$, $F_s = 0.76 \, \text{m}$, $A_s = \pi \times 50^2/4 = 1960 \, \text{mm}^2$ and $P_s = 300 \, \text{mm}$. For a DHP, $M_s = 50 \, \text{kg}$, $F_s = 0.50 \, \text{m}$, $A_s = \pi \times 43.7^2/4 = 1500 \, \text{mm}^2$ and $P_p = 100 \, \text{mm}$. Substituting these values into Equation (3.15) gives the equivalence:

SPT N-value (blows/300 mm) $\approx 2 \times$ DPH probe blows (blows/100 mm).

In clays, there is no real equivalence between probe blows and SPT blows. Probe results may however be used in conjunction with correlations with specific soil properties, as described in Chapters 5 and 6.

3.3.4 Static Cone Tests

Relative density may be estimated from cone resistance of the standard static cone, described in Appendix B. Figure 3.7 shows very general relationships between cone resistance and effective overburden pressure, compiled by Robertson and Campanella (1983). Robertson *et al.* (1983) also proposed a relationship between cone resistance and SPT N-values related to particle size, which can be seen in Figure 3.8.

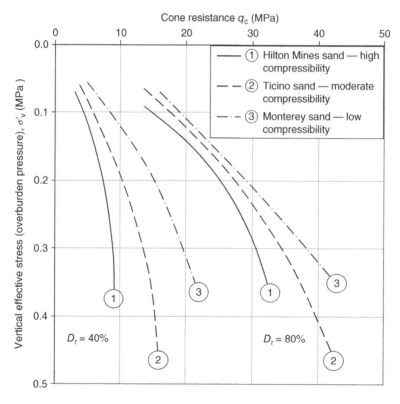

Figure 3.7 Relationships between cone resistance and vertical effective stress for different relative densities D_r. Adapted from Robertson and Campanella (1983). Reproduced with permission from NRC Research Press.

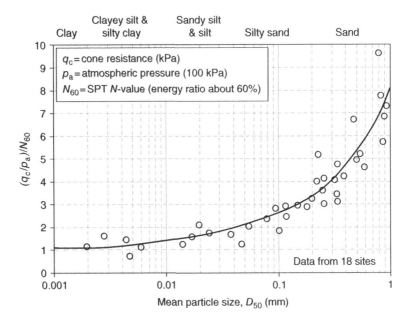

Figure 3.8 CPT–SPT correlation with grain size. Adapted from Robertson *et al.* (1983). Reproduced with permission of American Society of Civil Engineers.

References

AASHTO. 1982. *Standard specifications for transportation materials and methods of testing and sampling.* American Association of State Highway and Transportation Officials. Washington, DC, USA.

Al-Hussaini, M. M. and Townsend, F. C. 1975. Investigation of K 0 testing in cohesionless soils. Technical Report S-75-11, US Army Engineer Waterways Experimental Station, Vicksburg, Mississippi, 70 pp.

BS. 1990. *Methods of test for soils for civil engineering purposes.* British Standards Institution, BS 1377.

BS. 2011. *Geotechnical investigation and testing – Field testing.* British Standards Institute, BS EN ISO 22476-3: 2005 + A1.

Gibbs, H. J. and Holtz, W. G. 1957. Research on determining the density of sand by spoon penetration testing. *Proceedings of 4th International Conference on Soil Mechanics and Foundation Engineering,* London, I: 35–39.

Gregg, L. E. 1960. Earthworks. In: K. B. Woods (ed.) *Highway Engineering Handbook.* McGraw-Hill, New York.

Krebs, R. D. and Walker, E. D. 1971. *Highway Materials.* McGraw-Hill, New York.

Liao, S. C. and Whitman, R. V. 1986. Overburden correction factors for SPT in sand. *Journal of Geotechnical Engineering Division, ASCE* 112: 373–380.

Morin, W. J. and Todor, P. C. 1977. Laterites and lateritic soils and other problem soils of the tropics. United States Agency for International Development A lO/csd 3682.

Peck, R. B. and Bazaraa, A. R. S. S. 1969. Discussion. *Journal of Soil Mechanics and Foundation Division, ASCE* 95: 305–309.

Peck, R. B., Hanson, W. E. and Thornburn, T. H. 1974. *Foundation Engineering.* John Wiley, London, 514 pp.

Robertson, P. K. and Campanella, R. G. 1983. Interpretation of cone penetration tests, Part I: Sand. *Canadian Geotechnical Journal* 20: 718–733.

Robertson, P. K., Campanella, R. G. and Wrightman, A. 1983. SPT-CPT correlations. *Journal of Geotechnical Engineering, ASCE* 109: 1449–1459.

Seed, H. B. 1976. Evaluation of soil liquefaction effects on level ground during earthquakes. *ASCE Speciality Session, Liquefaction Problems in Geotechnical Engineering,* Preprint 2752, ASCE National Convention. Sept/Oct 1976, 1–105.

Skempton, A. W. 1986. Standard penetration test procedures and effects in sand of overburden pressure, relative density, particle size, ageing and over-consolidation. *Geotechnique* 36: 425–447.

Teng, W. C. 1962. *Foundation Design.* Prentice Hall, Englewood Cliffs, NJ.

Terzaghi, K. and Peck, R. B. 1967. *Soil Mechanics in Engineering Practice.* John Wiley, London, 729 pp.

Tokimatsu, K. and Yoshimi, Y. 1983. Empirical correlations of soil liquefaction based on SPT N-values and fines content. *Soils and Foundations* 23: 56–74.

Tomlinson, M. J. 1980. *Foundation Design and Construction.* Pitman, London, 793 pp.

Woods, K. B. and Litehiser, R.R. 1938. Soil mechanics applied to highway engineering in Ohio. *Ohio State University Engineering Experimental Station, Bulletin* 99.

4

Permeability

The coefficient of permeability is defined as the rate of flow through a unit area of soil under a unit pressure gradient. This assumes a linear relationship between the pressure gradient and rate of flow q (volume of flow per unit time), which is the basis for Darcy's law:

$$k = \frac{q}{Ai} \qquad (4.1)$$

where k is the coefficient of permeability;
  A is the area of flow; and
  i is the hydraulic pressure gradient.

If the rate of flow q is divided by the area A then the velocity of flow v is obtained and the permeability equation can be written:

$$k = \frac{v}{i}. \qquad (4.2)$$

From this, it can be seen that the coefficient of permeability can be thought of as the velocity of flow that results from a unit pressure gradient. Since pressure is usually measured as head of water and pressure is loss of

Soil Properties and their Correlations, Second Edition. Michael Carter and Stephen P. Bentley.
© 2016 John Wiley & Sons, Ltd. Published 2016 by John Wiley & Sons, Ltd.

head per unit distance, i typically has the dimensions m/m so that k has the units of velocity; typically m/s. However, it should be remembered that area A is the total area of soil being considered but part of this area will be occupied by solid particles so the area of flow will be less. This means that velocity v is only a notional value, used for calculating volumes of flow, and the true average velocity of flow v_t will be greater. With reference to the model soil sample illustrated in Figure 3.1:

$$v_t = v \times \frac{1+e}{e} = \frac{v}{n} \qquad (4.3)$$

where e and n are the voids ratio and porosity of the soil, respectively.

4.1 Effects of Soil Macro-Structure

The overall permeability of a soil mass is strongly influenced by its macro-structure; clays containing fissures or continuous bands of sand will have permeabilities that are many times that of the clay material itself. Also, since flow tends to follow the line of least resistance, stratified soils often have horizontal permeabilities which are many times the vertical permeability so that the effective overall permeability will be approximately equal to the horizontal permeability. Because of the small size of laboratory specimens and the way they are obtained and prepared, large-scale features are absent and test results do not give a true indication of field values in soils with a pronounced macro-structure. Moreover, laboratory tests usually constrain water to flow vertically through the specimen whereas the horizontal permeability may be the predominant factor for the soil mass. Field tests overcome these shortcomings, but, since the pattern of water flow from a well can only be guessed, interpretation of the test results is difficult and uncertain. One set of problems is therefore exchanged for another, as discussed in Chapter 1.

4.2 Typical Values

The typical range of values encountered is indicated in Figure 4.1 which is based on information originally presented by Casagrande and Fadum (1940). Superimposed on the chart are typical values of soil classes using the Unified classification system, for soils compacted to the heavy compaction standard (AASHTO 1982, T-180 (10 lb rammer) or BS 1990, 1377 Test 13 (4.5 kg rammer)). Typical permeability values for highway materials suggested by Krebs and Walker (1971) are given in Table 4.1.

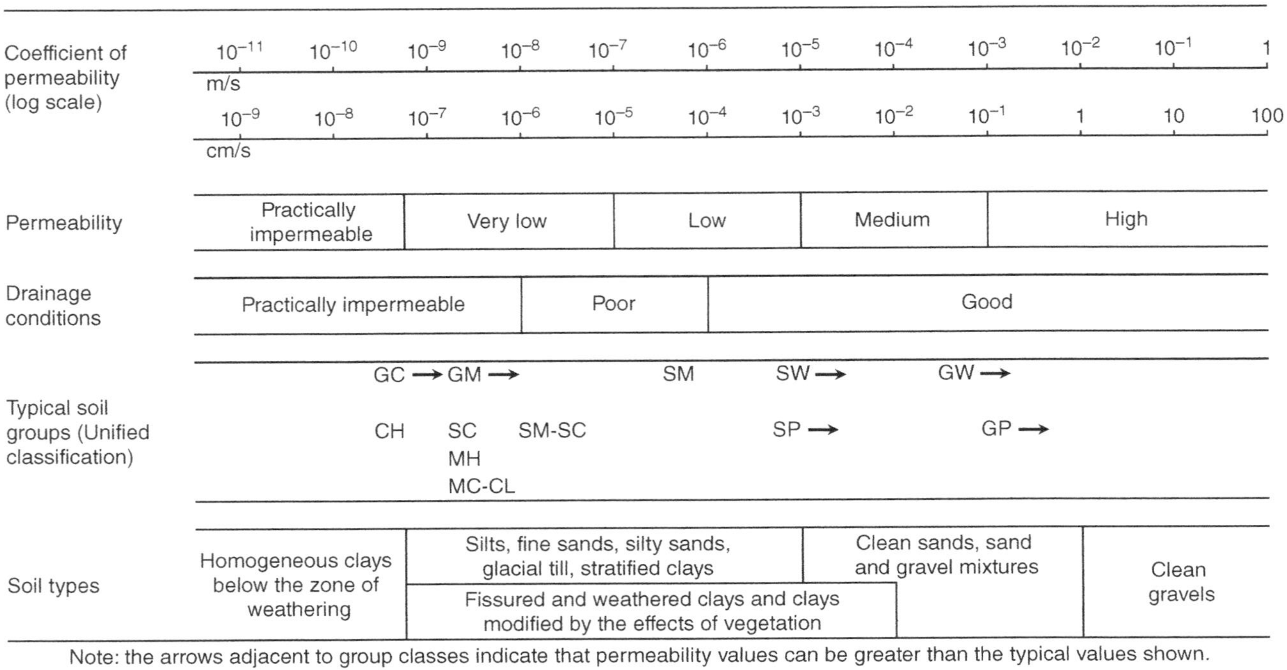

Coefficient of permeability (log scale)

m/s: 10^{-11} 10^{-10} 10^{-9} 10^{-8} 10^{-7} 10^{-6} 10^{-5} 10^{-4} 10^{-3} 10^{-2} 10^{-1} 1

cm/s: 10^{-9} 10^{-8} 10^{-7} 10^{-6} 10^{-5} 10^{-4} 10^{-3} 10^{-2} 10^{-1} 1 10 100

Permeability: Practically impermeable | Very low | Low | Medium | High

Drainage conditions: Practically impermeable | Poor | Good

Typical soil groups (Unified classification): GC → GM → | SM | SW → | GW → | SC | SM-SC | SP → | GP → | CH | MH | MC-CL

Soil types: Homogeneous clays below the zone of weathering | Silts, fine sands, silty sands, glacial till, stratified clays | Clean sands, sand and gravel mixtures | Clean gravels | Fissured and weathered clays and clays modified by the effects of vegetation

Figure 4.1 Typical permeability values for soils. Adapted from Casagrande and Fadum (1940).

Note: the arrows adjacent to group classes indicate that permeability values can be greater than the typical values shown.

Table 4.1 Typical values of permeability. Adapted from Krebs and Walker (1971).

Material	Permeability (m/s)
Uniformly graded coarse aggregate	0.4×10^{-3} to 4×10^{-3}
Well-graded aggregate without fines	4×10^{-3} to 4×10^{-5}
Concrete sand, low dust content	7×10^{-4} to 7×10^{-6}
Concrete sand, high dust content	7×10^{-6} to 7×10^{-8}
Silty and clayey sands	10^{-7} to 10^{-9}
Compacted silt	7×10^{-8} to 7×10^{-10}
Compacted clay	$<10^{-9}$
Bituminous concrete* (called asphalt in UK)	4×10^{-5} to 4×10^{-8}
Portland cement concrete	$<10^{-10}$

* New pavements; values as low as 10^{-10} have been reported for sealed, traffic-compacted highway pavement.

4.3 Permeability and Grading

A theoretical equation relating the coefficient of permeability to the soil and permeant properties was developed by Taylor (1948). This gave:

$$k = D_s^2 \frac{\gamma_p}{\mu} \frac{e^3}{(1+e)} c \qquad (4.4)$$

where k is the coefficient of permeability;

D_s is some effective particle diameter;

γ_p is the weight density (unit weight) of the permeant;

μ is the viscosity of the permeant;

e is the voids ratio; and

c is a shape factor.

In soils, the permeant is usually water and the effective particle diameter D_s is usually taken as the 10% (or effective) particle size D_{10}. This led to the Hazen formula:

$$k = C_1 D_{10}^2 \qquad (4.5)$$

where the constant C_1 replaces

$$\frac{\gamma_p}{\mu} \frac{e^3}{(1+e)} c. \qquad (4.6)$$

Based on experimental work with clean sands, Hazen (1911) proposed a value of between 0.01 and 0.015 for C_1, where k is in m/s and D_{10} is in mm.

However, this ignores the large effect that even small changes in e will have on the value of k, as can be seen from Taylor's equation, and can be expected to give only very approximate results. For instance, experimental work by Lambe and Whitman (1979) gives C_l values of between 0.01 and 0.42 with an average value of 0.16, while Holtz and Kovacs (1981) suggest a range of 0.004 to 0.12 with an average value of 0.01. The equation is usually considered to be valid for soils having a coefficient of permeability of at least 10^{-5} m/s, implying that it may be used for sand and gravel but is unreliable for silt and clay.

Figure 4.2 gives plots of k against D_{10}, based on experimental results, in which the value of e has been taken into account. It will be noted that the

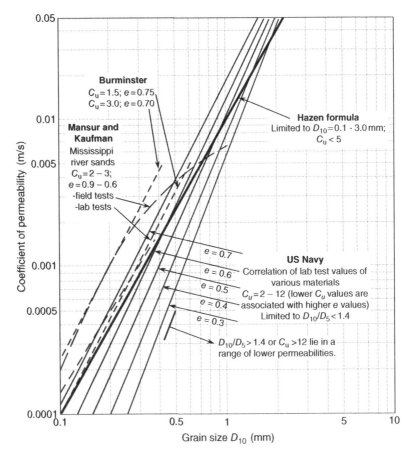

Figure 4.2 Permeability of sands and gravels. Adapted from US Naval Publications and Forms Center, US Navy (1982).

correlations given all relate to sands and gavels. Again, the greater range of particle size which is present in most clays and the effects of the clay mineralogy make such correlations more restrictive for clays.

It will be seen from the above that permeability estimates based on Hazen's formula should be regarded as very approximate at best, and possibly misleading in the case of silts and clays. The popular use of this formula to assess the spread of contaminants through soil should be discouraged except for preliminary estimates, and assessments should be based on actual (preferably field) measurements. Contaminant spread is also influenced by groundwater flows, which need to be measured and allowed for in assessments of the spread of contamination.

References

AASHTO. 1982. *Standard specifications for transportation materials and methods of testing and sampling*. American Association of State Highway and Transportation Officials. Washington, DC, USA.

Burmister, D. M. 1954. Principles of permeability testing of soils, symposium on permeability of soils. *ASTM Special Technical Publication* 163: 3–26.

BS. 1990. *Methods of test for soils for civil engineering purposes*. British Standards Institution, BS 1377.

Casagrande, A. and Fadum, R. E. 1940. *Notes on Soil Testing for Engineering Purposes*. Soil Mechanics Series No. 8, Harvard Graduate School of Engineering.

Hazen, A. 1911. Discussion of dams on sand foundations. *Transactions ASCE* 73: 199–203.

Holtz, R. D. and Kovacs, W. D. 1981. *An Introduction to Geotechnical Engineering*. Prentice-Hall, New Jersey, 733 pp.

Krebs, R. D. and Walker, E. D. 1971. *Highway Materials*. McGraw-Hill. Publication 272, Department of Civil Engineering, Massachusetts Institute of Technology, 107 pp.

Lambe, T. W. and Whitman, R. V. 1979. *Soil Mechanics*. John Wiley, New York, 55 pp.

Mansur, C. I. and Kaufman, R. I. 1962. Dewatering. In G. A. Leonards (ed.), *Foundation Engineering*. McGraw-Hill, New York, 241–350.

Taylor, D. W. 1948. *Fundamentals of Soil Mechanics*. John Wiley, New York, 700 pp.

US Navy. 1982. *Design Manual: Soil Mechanics, Foundations and Earth Structures*. Navy Facilities Engineering Command, Navfac, US Naval Publications and Forms Center.

5

Consolidation and Settlement

The settlement of soils in response to loading can be broadly divided into two types: elastic settlement and time-dependent settlement. Elastic settlements are the simplest to deal with; they are instantaneous, recoverable, and can be calculated from linear elastic theory. Time-dependent settlements occur in both granular and cohesive soils, although the response time for granular soils is usually short. In addition to being time-dependent, their response to loading is nonlinear and deformations are only partially recoverable. In clays, two types of time-dependent settlement are recognised.

- *Primary consolidation* results from the squeezing out of water from the soil voids under the influence of excess pore-water pressures generated by the applied loading. This can take place over many months or years in clays but is usually quick in sands and gravels due to their greater permeability.
- *Secondary compression* in clays and creep in sands occurs essentially after all the excess pore pressures have been dissipated, that is, after primary consolidation is substantially complete, but the mechanisms involved are not fully understood.

Settlements of granular soils, both elastic and creep movements, are more difficult to predict with any accuracy, largely because of the difficulty of

Soil Properties and their Correlations, Second Edition. Michael Carter and Stephen P. Bentley.
© 2016 John Wiley & Sons, Ltd. Published 2016 by John Wiley & Sons, Ltd.

obtaining and testing undisturbed soil samples, and settlements are usually estimated by indirect methods. Alternatively, plate bearing tests may be used but their results are difficult to interpret.

5.1 Compressibility of Clays

The compressibility of clays is usually measured by means of oedometer (consolidometer) tests or similar methods, as described in Chapter 1. Results may be expressed in a number of ways, leading to a sometimes confusing variety of compressibility parameters. As indicated in Figure 5.1, either sample thickness h or voids ratio e may be plotted against consolidation pressure p, which may itself be plotted either to a natural scale or, more usually, to a logarithmic scale.

Figure 5.1 also indicates how the previous maximum consolidation pressure may be estimated. Where this is greater than the current effective overburden pressure, the soil is overconsolidated to an extent defined by the overconsolidation ratio, OCR, where:

$$\text{OCR} = \frac{\text{previous maximum effective vertical pressure}}{\text{current effective overburden pressure}}. \tag{5.1}$$

Overconsolidation can occur in one of two ways: by a history of erosion where former overburden has been removed; or by desiccation, which cause effective stresses within the soil that have a similar consolidating effect to additional overburden.

5.1.1 Compressibility Parameters

The process of compression on a soil can be usefully illustrated by means of the model soil sample, shown in Figure 5.2. Recognising that compression takes place by a reduction in the volume of voids, with virtually no change in the volume of the solid particles, compressibility was originally defined by the coefficient of compressibility a_v which is the change in voids ratio per unit increase in pressure. In terms of the model soil sample,

$$a_v = -\frac{de}{dp} = \frac{e_1 - e_2}{p_2 - p_1} \tag{5.2}$$

and is the slope of the curve shown in Figure 5.1a when e is plotted against p.

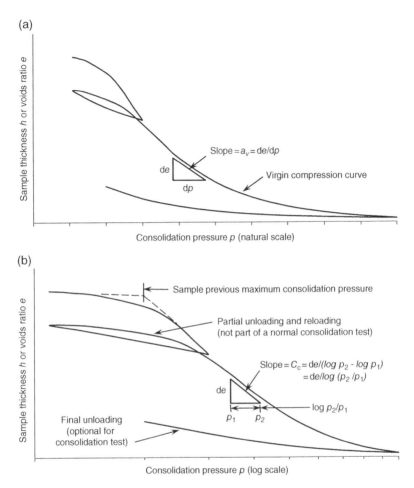

Figure 5.1   Basic features of plots of compressibility test results.

From an engineering viewpoint, it is the proportional change of thickness of a specimen that is of direct concern. For a constant cross-sectional area, this is proportional to the change of volume of a soil, and gives rise to the concept of the coefficient of volume of compressibility m_v which is much more commonly used:

$$m_v = -\frac{d(\text{volume})}{\text{volume}}\frac{1}{dp} = -\frac{dh}{h}\frac{1}{dp} = \frac{h_1 - h_2}{h_1}\frac{1}{(p_2 - p_1)}. \tag{5.3}$$

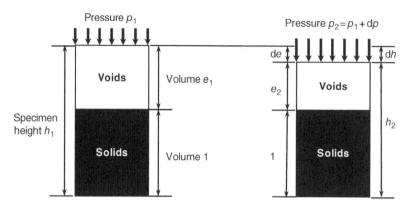

Figure 5.2   The compression process in terms of the model soil sample.

Referring to the model soil sample, m_v can also be expressed in terms of the voids ratio:

$$m_v = -\frac{de}{1+e_1}\frac{1}{dp} = \frac{e_1-e_2}{1+e_1}\frac{1}{(p_2-p_1)}. \qquad (5.4)$$

Comparing this second definition of m_v with the definition for a_v gives a relationship between the two compressibility parameters:

$$a_v = m_v(1+e). \qquad (5.5)$$

Because m_v is of more direct usefulness than a_v for settlement calculations, a_v is rarely used and m_v is often referred to simply as the 'coefficient of compressibility', although this is strictly incorrect.

It can be seen that the slope of the curve in Figure 5.1a is not constant. This means that the coefficients a_v and m_v also vary and that a given value applies only to a specific pressure range. However, the curve obtained in Figure 5.1b, when the logarithm of consolidation pressure is used, approximates much more closely to a straight line, at least on the virgin compression curve. This gives rise to a further measure of compressibility for the virgin compression part of the curve: the compression index C_c, which is similar in concept to the coefficient of compressibility a_v except that it uses the logarithm of pressure so represents the slope of the virgin compression curve in Figure 5.1b. That is:

$$C_c = -\frac{de}{d(\log p)} = \frac{e_1-e_2}{\log p_2 - \log p_1} = \frac{e_1-e_2}{\log \dfrac{p_2}{p_1}} \qquad (5.6)$$

where logarithms are taken to the base 10.

Comparing C_c with a_v in Equations (5.2) and (5.6) gives:

$$a_v = \frac{e_1 - e_2}{p_2 - p_1} C_c \frac{\log \dfrac{p_2}{p_1}}{e_1 - e_2} = C_c \frac{\log \dfrac{p_2}{p_1}}{p_2 - p_1}. \tag{5.7}$$

Now comparing a_v with m_v (Equation 5.5) gives:

$$m_v = \frac{a_v}{(1 + e_1)} = C_c \frac{\log \dfrac{p_2}{p_1}}{(1 + e_1)(p_2 - p_1)}. \tag{5.8}$$

For the recompression part of the curve, the recompression index C_r is used, defined in the same way as C_c but for recompression.

5.1.2   Settlement Calculations Using Consolidation Theory

Returning to the basic definition for the coefficient of volume compressibility,

$$m_v = -\frac{dh}{h} \frac{1}{dp} \tag{5.9}$$

it can be seen that, once m_v is known for a particular pressure range, the compression dh of a layer of thickness h due to a load increment dp can be calculated by simply rearranging to yield:

$$dh = m_v h \, dp. \tag{5.10}$$

Since dh is normally thought of as settlement ρ, and dp is the applied pressure σ, this becomes:

$$\rho = m_v H \sigma \tag{5.11}$$

where specimen thickness h is now replaced by the thickness H of the compressible stratum.

The average value of σ across a compressible layer due to some applied loading is usually calculated from elastic theory. Although not strictly valid for soils, which are not linear elastic materials, elastic theory gives sufficiently accurate values of stress distribution. Settlement can then be calculated using Equation (5.11). For very thick compressible layers, where the values of both the applied load and the coefficient of volume compressibility will vary significantly over depth, the calculation is usually carried out for a series of sub-layers.

Where values of C_c are obtained, m_v values may be calculated from them for the appropriate pressure ranges, and the settlement calculation carried out as described above. Alternatively, calculations may be carried out directly using C_c, with the appropriate substitutions for applied stress and layer thickness:

$$\rho = m_v H\sigma = C_c \frac{\log \dfrac{p_2}{p_1}}{(1+e_1)(p_2-p_1)} H(p_2 - p_1) \qquad (5.12)$$

giving

$$\rho = m_v H\sigma = C_c \frac{\log \dfrac{p_2}{p_1}}{(1+e_1)} H. \qquad (5.13)$$

It should be noted that, while only the change in applied pressure is used in the settlement calculation when using m_v, the actual initial and final pressures, including effective overburden pressure, must be used when using C_c.

5.1.3   Settlement Calculations Using Elastic Theory

Both stresses and displacements within a soil mass can be calculated using elastic theory; numerous solutions exist, covering a wide range of situations, many of which have been presented by Poulos and Davis (1974). Elastic solutions can be used in two ways when calculating settlement:

• to calculate profiles of vertical pressure within the ground, which can then be used in normal settlement calculations using consolidation theory as described above; and
• to calculate ground displacements directly, using the appropriate formulae for displacements within an elastic continuum.

The second approach means that settlements are calculated directly, without the intermediate stage of calculating stresses which are then used to calculate displacements using average values of the coefficient of volume compressibility. However, the problem with using elastic solutions to calculate settlements in this way is that it requires the evaluation of Young's modulus E and Poisson's ratio v, neither of which is measured, or is strictly meaningful, for soil consolidation problems.

Considering the basic equation for m_v, since the ratio dh/h can be thought of as a strain, m_v is strain/stress with units 1/stress; typically m²/kN or m²/MN. Thus, it is by definition akin to the reciprocal of Young's modulus E, and whereas E can be envisaged simplistically as the stress required to double the length of an object, m_v can be envisaged as an area of soil which, if subjected to a unit load, will just disappear. Of course, such absurdities do not occur in reality because the relationships are not valid for these extremes. Additionally, the relationship between E and m_v is not a simple reciprocal one because E is defined for a specimen with unrestrained sides, whereas m_v is defined for a specimen that is laterally constrained. The relationship between E and m_v therefore depends on the value of Poisson's ratio v, thus:

$$m_v = \frac{1}{E} \frac{(1+v)(1-2v)}{(1-v)}. \tag{5.14}$$

This relationship can then be used when calculating settlements using elastic theory. When used in this context, E is not strictly an elastic constant but it does represent the response of the soil to a single loading applied over a long period. To emphasise the point, the term 'deformation modulus' is sometimes used for E defined in this way. Elastic theory can therefore be used to calculate consolidation settlements, even though these are not really elastic (i.e. recoverable). The main problem lies in obtaining a value of Poisson's ratio that properly represents the consolidation behaviour of soils. Poisson's ratio is not measured in standard soil testing and, indeed, it is extremely difficult or impossible to obtain realistic measurements. However, it has been pointed out by Skempton and Bjerrum (1957) that very little lateral strain occurs during the consolidation of clays so that, effectively, Poisson's ratio is zero, giving:

$$E = \frac{1}{m_v} = M \tag{5.15}$$

where M is the deformation modulus or constrained modulus.

Another reason for choosing a zero value for Poisson's ratio is that calculated settlements based on elastic solutions then become identical to those based on consolidation theory, which has been shown over the years to give reasonable predictions provided suitable corrections are made for the pore-pressure response of the soil (Skempton and Bjerrum 1957).

5.1.4 Typical Values and Correlations of Compressibility Coefficients

Typical values of the coefficient of volume compressibility m_v are indicated in Table 5.1, along with descriptive terms for the various ranges of compressibility. Although m_v is the most suitable and most popular of the compressibility coefficients for the direct calculation of settlements, its variability with confining pressure makes it less useful when quoting typical compressibility values or when correlating compressibility with some other property. For this reason, the compression index C_c is sometimes preferred. Some tentative values of compression index, suggested by the authors, are also given in Table 5.1.

Skempton (1944) proposed the following relationship between compression index and liquid limit (LL) for normally consolidated clays:

$$C_c = 0.007(LL - 10).$$ (5.16)

Terzaghi and Peck (1967) proposed a similar relationship, based on research with clays of low and medium sensitivity:

$$C_c = 0.009(LL - 10).$$ (5.17)

This relationship is reported to have a reliability range of ±30% and to be valid for inorganic clays of sensitivity up to 4 (see Chapter 6) and a liquid limit up to 100.

In the authors' experience, the above correlations tend to overestimate the value of C_c (and hence m_v) and should preferably be used, if at all, in conjunction with some direct measurements of compressibility so that the correlations can be checked for reliability and, if appropriate, adjusted to suit local site conditions.

Based on the work of Skempton and Northey (1952) and Roscoe et al. (1958), Wroth and Wood (1978) used critical state soil mechanics considerations to deduce a relationship between compression index and plasticity index (PI) for remoulded clays:

$$C_c = \frac{1}{2} PI G_s$$ (5.18)

where G_s is the specific gravity of the soil solids.

Table 5.1 Typical values of the coefficient of volume compressibility m_v and the descriptive terms used.

Type of clay	Descriptive term[a]	Coefficient of volume compressibility, m_v (m²/MN)	Compression index, C_c[b]
Hard, heavily overconsolidated Glacial Till (Boulder Clay), stiff weathered rocks (e.g. completely weathered mudstone) and hard clays	Very low compressibility	<0.05	0.025
Stiff Glacial Till (Boulder Clay), marls, very stiff tropical residual clays	Low compressibility	0.05–0.1	0.025–0.05
Firm clays, glacial outwash clays, consolidated lake deposits, weathered marls, firm glacial till, normally consolidated clays at depth, firm tropical residual clays	Medium compressibility	0.1–0.3	0.05–0.15
Poorly consolidated alluvial clays such as estuarine deposits, and sensitive clays	High compressibility	0.3–1.5	0.15–0.75
Highly organic alluvial clays and peats	Very high compressibility	>1.5	0.75–5+

[a] Related to the coefficient of volume compressibility m_v.
[b] Based on an initial voids ratio of 0.5 and initial and final pressures of 100 kPa and 200 kPa, respectively.

Table 5.2, produced by Azzouz *et al.* (1976), gives a summary of a number of published correlations. Typical values of the recompression index C_r range from 0.015 to 0.35 (Roscoe *et al.* 1958), and are often assumed to be 5-10% of C_c.

Since none of the above correlations and typical values take into account the state of compaction of the soil, the resulting estimates of C_c and m_v are inevitably subject to significant errors. A relationship that takes into account the state of packing of a soil, as measured by the standard penetration test,

Table 5.2   Some published correlations for compression index C_c. Reproduced from Azzouz *et al.* (1976).

Correlation	Region of applicability
$C_c = 0.007(LL - 7)$	Remoulded clays
$C_c = 17.66 \times 10^{-5} m^2 + 5.93 \times 10^{-3} m - 0.135$	Chicago clays
$C_c = 1.15 (e_0 - 0.35)$	All clays
$C_c = 0.3 (e_0 - 0.27)$	Inorganic cohesive soil: silt, some clay; silty clay; clay
$C_c = 1.15 \times 10^{-2} m$	Organic soils: meadow materials, peats and inorganic silt and clay
$C_c = 0.75 (e_0 - 0.5)$	All clays
$C_c = 0.01 m$	Chicago clays

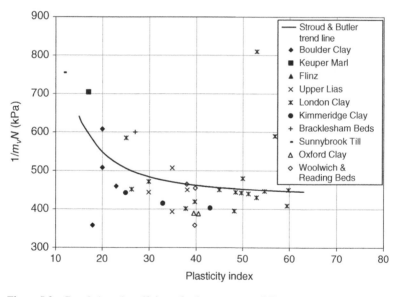

Figure 5.3   Correlation of coefficient of volume compressibility m_v with plasticity index and SPT N-value. Adapted from Stroud and Butler (1975). Reproduced with permission of Midland Geotechnical Society.

is given by Stroud and Butler (1975) for overconsolidated clays, stiff insensitive clays and soft rocks. Their correlation was presented graphically as shown in Figure 5.3. However, for use with spreadsheets it is convenient to express it as a mathematical relationship, and the authors have developed the relationship:

$$\frac{1}{m_v N} = \frac{56000}{(PI)^{2.07}} + 435. \tag{5.19}$$

This relationship is widely used for both overconsolidated and normally consolidated soils but it should be remembered that it was obtained specifically for overconsolidated soils, and is unlikely to be valid for all soil types. The mis-use of the relationship in this way is discussed by Reid and Taylor (2010). Ideally, it should be checked against consolidation test results for the specific site and soil type, then used to extend the quantity of results from SPT N-values once the correlation has been validated.

5.1.5   Settlement Corrections

If the results of oedometer tests are used directly to calculate settlements, the values obtained tend to overestimate the settlements that actually occur, particularly with overconsolidated clays. An exception to this is in the case of very sensitive clays, where predicted settlements may slightly underestimate actual values. The reason for this is that the pore-pressure response of clays in the field differs from that of confined laboratory specimens. This has been discussed by Skempton and Bjerrum (1957), who show that the ratio of actual settlement to calculated settlement depends on both the response of the pore-water pressures to applied loads and the geometry of each problem. The response of the pore-water pressures to loading can be measured in the triaxial test and is expressed in terms of Skempton's (1954) pore-pressure parameters, A and B. For saturated clays, actual settlement ρ_{field} is given by:

$$\rho_{field} = \mu \rho \tag{5.20}$$

where ρ is the calculated oedometer settlement and μ is a factor which depends on the pore-pressure parameter.

The distribution of stresses across a layer of soil depends on the ratio of width b of a foundation to thickness H of the layer. Values of μ can be obtained for given values of pore-pressure parameter A from Figure 5.4.

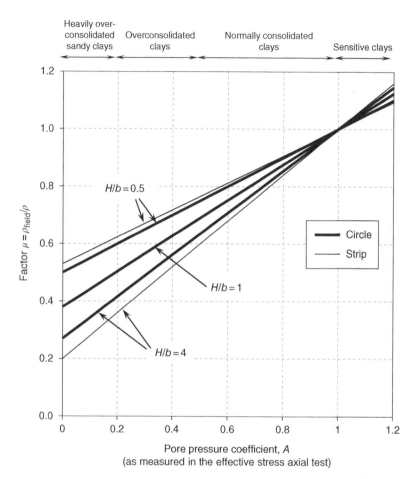

Figure 5.4 Typical values of factor μ for a foundation of width b on a compressible layer of thickness H. Adapted from Skempton (1954). Reproduced with permission of ICE.

Values of parameter A are not normally measured in the laboratory tests commonly used for foundation design but they are found to depend on the consolidation history of the clay, particularly the degree of overconsolidation. For most practical purposes it is sufficient to use values of μ selected from Figure 5.5.

Type of clay	Factor μ		
	$H/b = 0.5$	$H/b = 1$	$H/b = 4$
Very sensitive clays (soft alluvial, estuarine, marine clays)	1.0–1.1	1.0–1.1	1.0–1.1
Normally consolidated clays	0.8–1.0	0.7–1.0	0.7–1.0
Overconsolidated clays (Lias, London, Oxford, Weald clays)	0.6–0.8	0.5–0.7	0.4–0.7
Heavily overconsolidated clays (Glacial Till, marl)	0.5–0.6	0.4–0.5	0.2–0.4

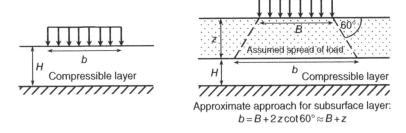

Figure 5.5 Typical values of consolidation factor μ for various types of soil. Adapted from Skempton (1954).

5.2   Rate of Consolidation of Clays

The rate of settlement of a saturated soil is expressed by the coefficient of consolidation, c_v. Theoretically, consolidation takes an infinitely long time to be completed and it is usual to calculate the time taken for a given degree of consolidation U to occur, where U is defined by:

$$U = \frac{\text{consolidation settlement after a given time } t}{\text{final consolidation settlement}}. \tag{5.21}$$

The time t for a given degree of consolidation to occur is given by:

$$t = \frac{T_v d^2}{c_v} \tag{5.22}$$

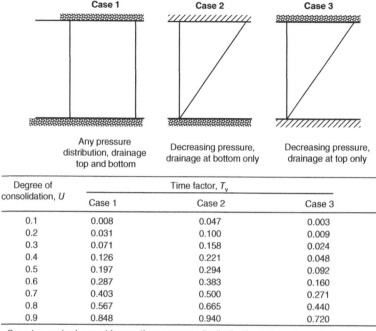

Case 1 may also be used for a uniform pressure distribution for drainage top or bottom only.

Figure 5.6 Typical values of time factor T_v

where d is the maximum length of the drainage path (equal to half the layer thickness for drainage top and bottom) and T_v is called the basic time factor. Values of T_v for various values of U are given in Figure 5.6.

The rate of settlement of a soil, and hence the value of c_v, is governed by two factors: the amount of water to be squeezed out of the soil; and the rate at which that water can flow out. The amount of water to be squeezed out depends on the coefficient of volume compressibility m_v, and the rate at which it will flow depends on the coefficient of permeability k. The relationship between c_v, m_v and k is:

$$c_v = \frac{k}{m_v \gamma_w} \tag{5.23}$$

where γ_w is the weight density (or unit weight) of water.

Because of the wide range of permeability that exists in soils, the coefficient of consolidation can itself vary widely from less than 1 m²/year for

Table 5.3 Typical values of the coefficient of consolidation c_v. Adapted from Holtz & Kovacs (1981).

Soil type (Unified classification)	c_v	
	cm²/s	m²/a
Low plasticity clays (CL)	0.001–0.006	3–19
Low plasticity glacial lake clays (CL)	0.0006–0.0009	2–3
Low plasticity mud (CL)	0.0002–0.0004	0.6 – 1.2
Medium plasticity clays (CL–CH)	<0.0001–0.0003	<0.3–0.9
Volcanic silt (MH)	0.0001–0.00015	0.3–0.5
Organic silt (OL)	0.0002–0.001	0.6–3.0

clays of low permeability to 1000 m²/year or more for very sandy days, fissured clays and weathered rocks. Some typical values for clays are given in Table 5.3. Attempts have been made to correlate c_v with other properties, principally liquid limit. That c_v is related to clay type, as measured by liquid limit, seems reasonable, but it is also related to other factors, especially the state of compaction of the soil, so that correlations for c_v can be no more than tentative, giving rough indications of the likely range for any soil. One such correlation, given by the US Navy (1982), is shown in Figure 5.7.

In addition to the originally published trend lines shown in the figure, dashed lines also indicate approximations to the correlations using mathematical functions, which may be more useful than the graphical presentation for use with spreadsheets. These functions return values within 12% of the original graphical values given by the US Navy's trend lines.

5.3 Secondary Compression

Secondary compression is a volume change under load that takes place at constant effective stress; that is, after the excess pore-water pressure has dissipated. It is thought to result from compression of the constituent soil particles at a microscopic or molecular scale and is particularly significant in organic soils and estuarine deposits. Coefficients of secondary compression may be defined in a way that is analogous to the definitions of compression index and modified compression index, except that the indices are related to time instead of pressure. The equivalent of a_v is therefore the secondary compression index C_α:

$$C_\alpha = -\frac{de}{d(\log t)} \tag{5.24}$$

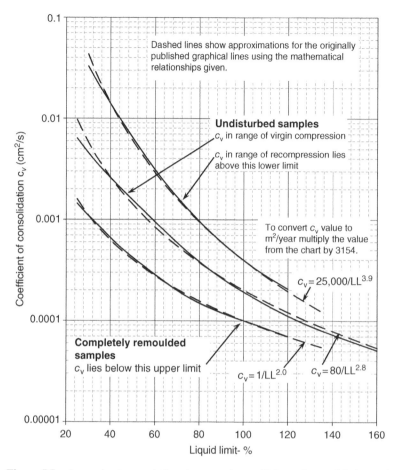

Figure 5.7 Approximate correlations between the coefficient of consolidation and liquid limit. Adapted from US Naval Publications and Forms Center, US Navy (1982).

where de is the change in voids ratio over a time interval dt, from time t_1 to time t_2 (see Fig. 5.8).

Similarly, the equivalent of m_v is the modified secondary compression index $C_{\alpha\varepsilon}$ (sometimes referred to as the secondary compression index and sometimes given the symbol C'_α):

$$C_{\alpha\varepsilon} = -\frac{dh/h}{d(\log t)} = \frac{C_\alpha}{(1+e)} \tag{5.25}$$

where e is the voids ratio at the start of the linear portion of the e-log p (or dh-log p) curve.

Secondary compression calculations are carried out by rearranging Equations (5.24) and (5.25). Specimen compression dh becomes secondary compression settlement ρ_c; specimen thickness h becomes layer thickness H; and time is taken over an interval from t_1 to t_2:

$$\rho_c = C_{\alpha\varepsilon} H \log\left\{\frac{t_2}{t_1}\right\} = \frac{C_\alpha}{1+e} H \log\left\{\frac{t_2}{t_1}\right\}. \tag{5.26}$$

For the purpose of secondary settlement calculations, secondary compression settlement is assumed to start when primary settlement is substantially complete. Thus, if primary settlements were substantially complete in 6 months, the value of t_1 would be 6 months. The value of t_2 depends on the assumed lifespan of the structure under consideration.

Values of C_α or $C_{\alpha\varepsilon}$ are obtained from e-log p or dh-log p plots, as indicated in Figure 5.8. However, standard consolidation tests do not usually continue for long enough to obtain the secondary compression part of the curve, and values are often estimated using typical values or correlations. C_α is usually assumed to be related to C_c, with values of C_α/C_c typically in the range 0.025 to 0.06 for inorganic soils and 0.035 to 0.085 for organic soils. A list of some published values of the ratio C_α/C_c

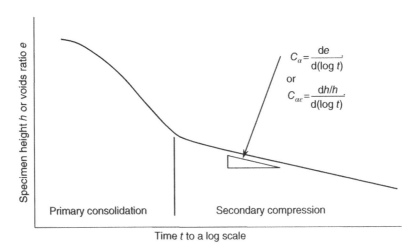

Figure 5.8 Plotting and calculation of secondary compression.

Table 5.4 Typical values of the ratio of secondary compression index to compression index. Adapted from Mesri and Godlewski (1977).

Soil	C_α/C_c
Organic silts (OL)	0.035–0.06
Amorphous and fibrous peat (Pt)	0.035–0.085
Canadian muskeg (Pt–OL)	0.09–0.10
Leda clay, Canada (CL)	0.03–0.06
Post-glacial Swedish clay (CL)	0.05–0.07
Soft blue clay, Victoria, BC (CL–CH)	0.026
Organic clays and silts (ML–MH)	0.04–0.06
Sensitive clay, Portland, ME (CL–CH)	0.025–0.055
San Francisco Bay mud (CL)	0.04–0.06
New Liskeard (Canada) varved clay (CL)	0.03–0.06
Mexico City clay (MH)	0.03–0.035
Hudson River silt (ML)	0.03–0.06
New Haven organic silt (OL)	0.04–0.075

is given in Table 5.4. Mesri (1973) obtained a relationship between $C_{\alpha\varepsilon}$ and liquid limit, given in Figure 5.9. This includes an indication of the original data points and ranges of values obtained by Mesri, from which it can be seen that there is a wide scatter of results. A further relationship, between $C_{\alpha\varepsilon}$, and moisture content, quoted by the US Navy, is given in Figure 5.10.

A shortcoming of secondary compression theory is that it does not take into account the value of the load that leads to the primary, then secondary compression. Thus, a single-storey building would, in theory, experience the same secondary compression as a six-storey building. This is clearly nonsense, and differences would actually be reflected in the value of C_α or $C_{\alpha\varepsilon}$ by obtaining it for the appropriate load range during testing. However, this is not the case where values are estimated from correlations or typical values, so such values should be viewed with scepticism. Nevertheless, the experience of one the authors using estimates of $C_{\alpha\varepsilon}$ based on Mesri's correlations for a large artificial island was that settlement estimates were obtained which reasonably matched values obtained by back-analysis during the course of monitoring over several years. Small modifications to the original estimates were then made to produce a match between actual and predicted settlement over the monitoring period, allowing settlements to be predicted with some confidence for years ahead.

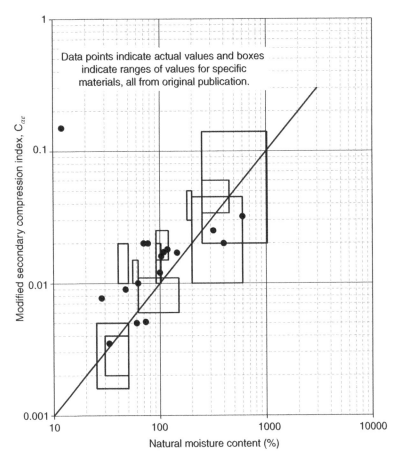

Figure 5.9 Correlation between modified secondary compression index and natural moisture content. Adapted from Mesri (1973). Reproduced with permission of American Society of Civil Engineers.

5.4 Settlement of Sands and Gravels

As mentioned in the introductory remarks to this chapter, the near-impossibility of obtaining and testing undisturbed samples of granular soils means that consolidation testing is not possible. Instead, settlements are usually estimated from *in situ* test results, most commonly using the standard penetration test and, increasingly, probes, in the form of static or dynamic cones. *In situ* plate bearing tests are also sometimes used to obtain settlement estimates, but their use is often limited by the cost of carrying out a sufficiently large-scale test to give meaningful results.

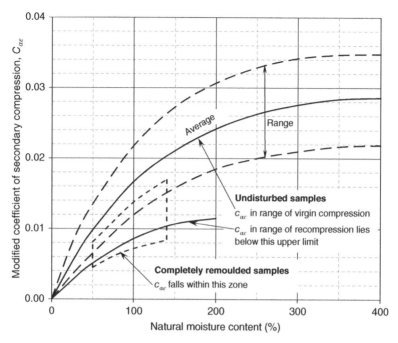

Figure 5.10 Approximate correlation between moisture content and the modified coefficient of secondary compression. Adapted from US Naval Publications and Forms Center, US Navy (1982).

5.4.1 Methods Based on Standard Penetration Tests

There are several methods for estimating settlements of structures on sands and gravels using SPT N-values. Because the standard penetration test does not measure the settlement characteristics of these materials the methods are generally empirical; based on observation of the behaviour of actual structures and comparing settlements with foundation size and the SPT N-value of the founding material.

Considering the practical problems of obtaining meaningful SPT results, especially in sands below the water table, and the disagreements over various corrections to be applied to the results, these correlations are of dubious value in many cases. Yet settlement estimates are often the controlling factor in determining allowable bearing pressures for granular soils, which generally have high ultimate bearing capacities for all except the narrowest foundations. In view of all these considerations, it is surprising that settlement calculations for granular soils have for so long relied on such an unsatisfactory procedure. Perhaps it reflects a lack of problems with foundations on granular soils compared with those founded on clay soils.

5.4.1.1 Traditional Methods

For many decades, the most commonly used correlation for settlement estimates in sands, based on SPT results, was that established by Terzaghi and
Peck (1967). Figure 5.11, which gives allowable bearing pressures for a
permissible settlement of 25 mm is based on Terzaghi and Peck's original
graph with values converted to SI units and with additional lines for
intermediate N-values values. Meyerhof (1956, 1974) also produced

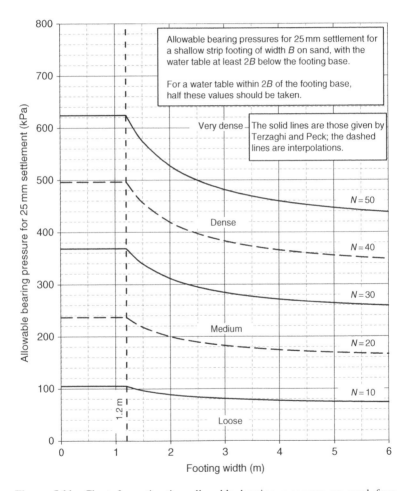

Figure 5.11 Chart for estimating allowable bearing pressures on sand from
standard penetration test results. Adapted from Terzaghi and Peck (1967).

relationships between SPT results and settlement which gave similar values to those of Terzaghi and Peck. However, Terzaghi and Peck point out that the correlations show wide scatter and should not be regarded as anything more than a rough-and-ready guide.

For foundations on saturated sand, Terzaghi and Peck suggest that allowable bearing pressures should be reduced to half of those given in their chart for shallow foundations and to two-thirds the chart values where the founding depth is about the same as the footing width, or deeper.

For permissible settlements other than the 25 mm assumed in the chart, allowable bearing pressures should be adjusted pro-rata.

Both the Meyerhof and the Terzaghi and Peck values have been criticised as being too conservative, and Bowles (1982) suggests that, in the light of field observations and the stated opinions of many authors, the Meyerhof equations should be adjusted to give an approximate 50% increase in allowable bearing pressure q_a. Bowles proposes:

$$q_a = \frac{N}{0.05} K_d \qquad (5.27)$$

for footing widths up to 1.2 m and

$$q_a = \frac{N}{0.08} \left\{ \frac{B+0.3}{B} \right\}^2 K_d \qquad (5.28)$$

for footing widths B exceeding 1.2 m;

where q_a is the foundation pressure that would produce 25 mm of settlement;
N is the SPT N-value (blows per 300 mm);
K_d is a depth factor $(=1+0.33D/B$ up to a maximum value of 1.33); and
D is the depth of the foundation base in metres.

The resulting curves for the case of $D=0$ (a foundation at the surface) are given in Figure 5.12. For a subsurface footing founded at depth D, values may be multiplied by K_d as defined above.

Raft foundations are known to settle less than strip footings, and Tomlinson (1980) suggests that the allowable pressures obtained from Figure 5.11 be doubled for this type of foundation. Alternatively, Bowles gives a modified form of the Meyerhof equation for rafts:

$$q_a = \frac{N}{0.08} K_d. \qquad (5.29)$$

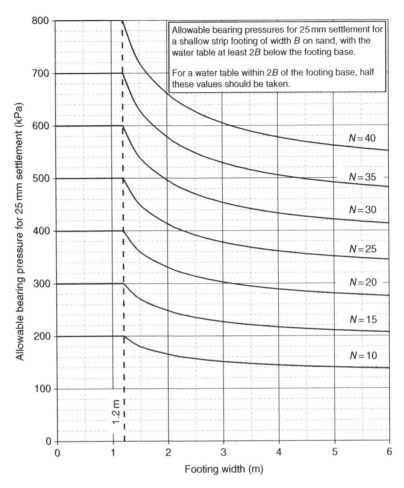

Figure 5.12 Chart for estimating allowable bearing pressures on sand from standard penetration test results based on recommendations by Bowles (1982).

It should be noted that none of these methods gives any indication of the rate of settlement. It is normally assumed that this will be rapid, with a high proportion of settlement taking place during the construction period. However, this assumption may be too optimistic in the case of large or settlement-sensitive structures on loose sands. The problem of rate of settlement is addressed by the Burland and Burbridge method, discussed in the following section.

5.4.1.2 The Burland and Burbridge Method

An alternative method of estimating settlements in sands and gravels has been given by Burland and Burbridge (1985) This is still based on SPT results and observation of full-scale buildings but it is a more comprehensive procedure, which takes into account foundation shape and layer thickness, and allows settlement predictions to be made for any number of years after construction. It also gives three estimates of settlement: a lower bound, a best estimate and an upper bound. Settlement of a footing of length L and width B (m) at time t after application of a load is given by the equation:

$$s = f_s f_1 f_t \left(q - \frac{2}{3} p \right) B^{0.7} I_c \qquad (5.30)$$

where

- f_s is a shape factor, given by:

$$f_s = \left\{ \frac{1.25 \dfrac{L}{B}}{\dfrac{L}{B} + 0.25} \right\}^2 \qquad (5.31)$$

- f_1 is a layer depth factor, given by $(H/z)(2 - H/z)$ provided $H < z$ and is equal to 1 when $H > z$, where H is the depth (m) of the gravel layer beneath the footing and z is the depth of influence of the footing (m), normally given by:

$$10^{0.74 \log B}$$

- f_t is a time factor, given by: $1 + R_3 + R \log(t/3)$ where t is time in years after application of the load for which settlement is to be calculated, and for static loads $R_3 = 0.3$ and $R = 0.2$.
- q is the foundation pressure (kPa).
- p is the previous maximum effective overburden pressure (kPa). (Note that from Equation (5.30) there will be no predicted settlement if the previous maximum overburden pressure is 1.5 times the foundation pressure, or more.)
- I_c is an influence factor, given by:

$$I_c = \frac{10^{a - b \log N}}{100} \qquad (5.32)$$

where $a = 2.2$ (best estimate); 1.778 (lower limit) or 2.98 (upper limit); $b = 1.4$ (best estimate); 1.31 (lower limit) or 1.68 (upper limit); and N is the SPT N-value.

This method probably represents a significant improvement on the older Terzaghi and Peck and Meyerhof methods, but the computation is much more complicated and is better performed using a spreadsheet.

The calculation can be greatly simplified by taking the cases of a square or strip footing on a deep layer of sand or gravel. In these cases, settlement is given by:

$$\rho = f_t q B^{0.7} I_c \tag{5.33}$$

for a square footing, and:

$$\rho = 1.56 f_t q B^{0.7} I_c \tag{5.34}$$

for a strip footing,

where $f_t = 1.0$ for immediate settlement;
       1.3 for settlement after 3 years;
       1.6 for settlement after 100 years;

and I_c can be obtained from the SPT N-value using Figure 5.13.

It can be seen that, in contrast to the traditional methods discussed in the previous section, no allowance is made for the depth of water table beneath the foundation base, although a water table above the foundation base may reduce the initial effective overburden pressure, increasing settlement predictions.

5.4.1.3   The Deformation Modulus Method

Work by Menzenbach (1967) established a rough relationship between deformation modulus M and SPT N-value, as shown in Figure 5.14. This can be used in conjunction with elasticity theory to obtain settlement predictions. For instance, for a strip foundation of width B and loading intensity q, settlement ρ is given by:

$$\rho = \frac{2.25\left(1 - \upsilon^2\right) q B}{M} \tag{5.35}$$

where Poisson's ratio υ is usually taken as 0.15 for sands.

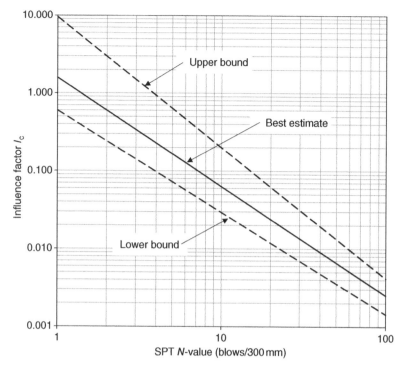

Figure 5.13 Correlation between influence factor I_c and SPT N-value for estimating settlements on sand using the method of Burland and Burbridge (1985).

Values of allowable bearing pressure for 25 mm settlement obtained in this way are broadly in line with those obtained from Figure 5.12.

Like the traditional settlement estimate methods discussed above, no indication is given of the rate of settlement or the amount that is likely to have occurred by the end of construction.

5.4.1.4   Use of N-Value Correction Factors when Estimating Settlement

It was discussed in Chapter 3 that, when estimating SPT N-values to assess relative density, correction factors may be applied as illustrated in Figure 3.6. However, the question remains as to whether such correction factors should be used when estimating settlement. Since settlement estimates based on N-values are empirical, the answer depends on whether the researchers who produced the settlement estimates applied such correction factors.

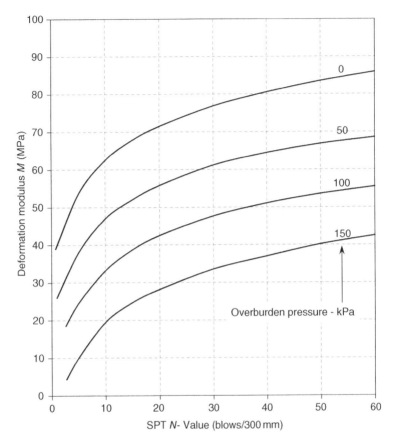

Figure 5.14 Correlation between deformation modulus M and SPT N-value for granular soils based on recommendations by Menzenbach (1967).

Terzaghi and Peck would not have applied such correction factors since their work predates the publications on N-value correction factors, as does the work of Menzenbach in producing correlations with deformation modulus. Since Bowles's factors are also based on the earlier works of Terzaghi and Peck and Meyerhof, it seems reasonable to assume that his recommendations also do not take account of correction factors. Burland and Burbridge's method, though published much later, also makes no mention of correction factors.

Based on these considerations, it seems unreasonable to apply such correction factors, which would result in higher recommendations of allowable bearing pressure, or much lower estimates of settlement. The authors therefore recommend that correction factors be not applied to settlement estimates based on SPT N-values.

5.4.2   Methods Based on Plate Bearing Tests

Plate bearing tests offer a more direct method of measuring settlements, but the usefulness of the results is limited by two constraints:

1. the depth of ground stressed by a plate is only a fraction of that stressed by a full-sized foundation; and
2. settlement predictions require knowledge of the scale effects between the settlement of a plate and that of a full-sized foundation, for both total settlement and the rate of settlement.

The most commonly used correlation for scale effects between plate and foundation settlements is that given by Terzaghi and Peck (1967):

$$\rho = \rho_1 \left\{ \frac{2B}{1+B} \right\} \qquad (5.36)$$

where ρ is the settlement of a square foundation of side B feet and ρ_1 is the settlement of a 1-foot square plate. If the foundation width is measured in metres, this becomes (for a 300 mm square plate):

$$\rho = \rho_1 \left\{ \frac{2B}{0.3+B} \right\}. \qquad (5.37)$$

An alternative, and more general, relationship was derived by Menard and Rousseau (1962):

$$\frac{\rho_1}{\rho_2} = \left\{ \frac{B_1}{B_2} \right\}^{\alpha} \qquad (5.38)$$

where ρ_1 and ρ_2 are the settlements of the plate and footing, respectively;
B_1 and B_2 are their respective widths; and
α is a factor that depends on the soil type. Typical values of α are:
     sands and gravels, $\frac{1}{3}$ to $\frac{1}{2}$
     saturated silts, $\frac{1}{2}$
     clays and dry silts, $\frac{1}{2}$ to $\frac{2}{3}$
     compacted fill, 1.

5.5   Assessment of Settlement Parameters from Static Cone Penetration Testing

5.5.1   Coefficient of Volume Compressibility

A rough approximation of the coefficient of volume compressibility m_v with static cone resistance can be obtained for normally consolidated clays and lightly overconsolidated clays and silts up to firm consistency using a correlation by Meigh (1987):

$$m_v = \frac{1}{\alpha q_c} \tag{5.39}$$

where α is a coefficient, between 2 and 8, which depends on the soil type, as indicated in Table 5.5;

and q_c is the measured cone resistance (pressure).

The relationship is valid up to a cone resistance q_c of 1.2 Mpa. The units of m_v will depend on the units of q_c; if q_c is in MPa (MN/m^2), m_v will be in m^2/MN.

Table 5.5   Typical values of coefficient α used to estimate m_v values from static cone penetration testing. Reproduced from Meigh (1987).

Soil type and classification (Unified system)	α (=1/($m_v q_c$))	
	Reference cone[a]	Mantle cone[b]
Highly plastic clays and silts (CH, MH)	2.5–7.5	2–6
Clays of intermediate or low plasticity (CI, CL)		
q_c < 700 kPa	3.7–10	3–8
q_c > 700 kPa	2.5–6.3	2–5
Silts of intermediate or low plasticity (MI, ML)	3.5–7.5	3–6
Organic silts (OL)	2.5–10	2–8
Peat and organic clay (Pt, OH)		
moisture content 50–100%	1.9–5.0	1.5–4.0
moisture content 100–200%	1.25–1.9	1.0–1.5
moisture content >200%	0.5–1.25	0.4–1.0

[a] The 10-tonne cone shown in Figure B7, Appendix B.
[b] Older mechanical cone in which the tip and sleeve are advanced separately, with similar tip and sleeve sizes to the reference cone (see Appendix B).

5.5.2   *Coefficient of Consolidation*

The coefficient of consolidation can be estimated from a dissipation test using a piezocone, in which the rate of dissipation of excess pore-water pressure at the cone tip is measured as described in Appendix B.

The time t_{50} taken for 50% excess pressure to dissipate is obtained from plots of excess pore pressure against the logarithm of time or the square root of time, as described in Appendix B. This is then used to estimate the coefficient of consolidation for horizontal water flow c_h using an established correlation, the most popular of which is probably that given by Robertson *et al.* (1992), shown in Figure 5.15a.

It will be noted, however, that the correlation depends on the value of the rigidity index I_r, defined as:

$$I_r = \frac{G}{S_u} \qquad (5.40)$$

where G is the shear modulus (normally taken as G_{50}, the shear modulus for 50% compressive strength) and S_u is the undrained shear strength.

However, there is debate about the best methods to measure these values relevant to the estimation of c_h, and the measurement itself is somewhat specialised. It is therefore convenient to estimate I_r values from correlations. A correlation between I_r and plasticity index for varying degrees of over-consolidation ratio (OCR), established by Keaveny (1985), is given in Figure 5.15b. Keaveny states that this gives I_r values to ±50%, but this may be optimistic for all soils. Alternative correlations have been published including one by Mayne (2001). Mayne states his correlations give slightly better estimates, but they use properties that require laboratory testing of samples which reduces the advantage of using cone testing; hence the earlier Keaveny correlation is presented here. The values given in Figure 5.15b can be obtained from the approximate correlation suggested by Mayne:

$$I_r = \frac{\exp\left\{\dfrac{137-\mathrm{PI}}{23}\right\}}{1+\ln\left\{1+\dfrac{\left(\mathrm{OCR}+1\right)^{3.2}}{26}\right\}^{0.8}} \qquad (5.41)$$

where PI is the plasticity index, OCR is the overconsolidation ratio and ln is $\log_e$.

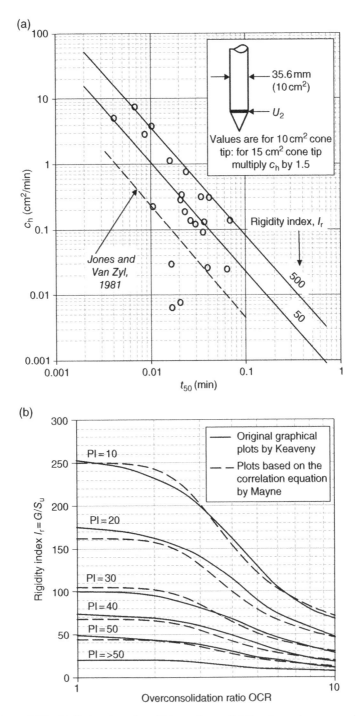

Figure 5.15 (a) Estimation of coefficient of consolidation from piezocone values. Adapted from Robertson *et al.* (1992). Reproduced with permission from NRC Research Press. (b) Estimation of rigidity index for use in chart (a). Adapted from Keaveny (1985) and Mayne (2001).

An alternative method of estimating I_r using only cone penetration testing (CPT) data, obviating the need for any sampling and laboratory testing, has been proposed by Krage *et al.* (2014). This requires a seismic cone which has an accelerometer to detect shear waves generated at the surface by a sledge hammer hitting a plate at the surface. This development is very recent and is so far based on only a relatively small number of trials, but may become a useful method in the future.

Discussions of the problems associated with predicting c_h values from static cone tests are given in a number of publications, including that by Broussard (2011). It will be seen from the above discussion that estimates of the coefficient of consolidation obtained from static cone testing should be viewed as only approximate. Given that the coefficient of compressibility is often a poor predictor of rates of consolidation, the use of field permeability measurements to obtain its value based on the relationship between these two parameters offers a more reliable, if more expensive, method of predicting settlement rates.

References

Azzouz, A. S., Krizek, R. J. and Corolis, R. B. 1976. Regression analysis of soil compressibility. *Soil and Foundations* 16(2): 19–29.

Bowles, J. E. 1982. *Foundation Analysis and Design.* McGraw-Hill, Singapore, 816 pp.

Broussard, N. S. 2011. Proper selection of rigidity index (Ir=G/Su) relating to CPT correlations. ECI 284 Term Project. University of California, Davis.

Burland, J. B. and Burbridge, M. C. 1985. Settlement of foundations on sand and gravel. *Proceedings of the Institution of Civil Engineers* 78(1): 1325–1381.

Holtz, R. D. and Kovacs, W. D. 1981. *An Introduction to Geotechnical Engineering.* Prentice-Hall, New Jersey, 733 pp.

Jones, G.A. and Van Zyl, D.J.A. 1981. The piezometer probe: a useful tool. *Proceedings of the 10th International Conference on Soil Mechanics and Foundation Engineering,* Stockholm, 2: 489–496.

Keaveny, J. 1985. *In-situ determination of drained and undrained shear strength using the cone penetration test.* Dissertation presented to the University of California, Berkeley, in partial fulfilment of the requirement for the degree of Doctor of Philosophy.

Krage, C. P., Broussard, N. S. and De Jong, J. T. 2014. Estimating rigidity index (I_R) based on CPT measurements. In: *Proceedings of 3rd International Symposium on Cone Penetration Testing,* Las Vegas.

Mayne, P. W. 2001. Stress-strain-strength-flow parameters from enhanced in-situ tests. In: *Proceedings of the International Conference on In-Situ Measurement of Soil Properties and Case Histories,* Bali.

Meigh A. C. 1987. *Cone Penetration Testing: Methods and Interpretation.* Butterworths, London.

Menard, L. and Rousseau, J. 1962. L'evaluation des tassements. Tendances nouvelles. *Sols Soils* 1: 13–20.

Menzenbach, E. 1967. Le capacidad suportante de pilotes y grupos de pilotes. *Technologia (Ingeneria Civil)* 2(1): 20–21.

Mesri, G. 1973. The coefficient of secondary compression. *Proceedings ASCE Journal of Soil Mechanics and Foundation Division* 99: 123–137.

Mesri, G. and Godlewski, P. M. 1977. Time and stress compressibility inter-relationships. *Proceedings ASCE Journal of Geotechnical Engineering Division* 103: 417–430.

Meyerhof, G. G. 1956. Penetration tests and bearing capacity of cohesionless soils. *Proceedings ASCE Journal of Soil Mechanics and Foundation Division* 82: 1–19.

Meyerhof, G. G. 1974. General report: outside Europe. In: *Proceedings of European Symposium on Penetration Testing*, Stockholm, 2: 40–48.

Poulos, H. G. and Davis, E. H. 1974. *Elastic Solutions for Soil and Rock Mechanics*. John Wiley, New York, 411 pp.

Reid, A. and Taylor, J. 2010. The misuse of SPTs in fine soils and the implications of Eurocode 7. *Ground Engineering* July 2010: 28–31.

Robertson, P. K., Sully, J. P., Woeller, D. J., Lunne, T., Powell, J. J. M. and Gillespie, D. G. 1992. Estimating the coefficient of consolidation from piezocone tests. *Canadian Geotechnical Journal* 29(2): 529–538.

Roscoe, K. H., Schofield, A. N. and Wroth, C. P. 1958. On the yielding of soils. *Geotechnique* 8: 2–52.

Skempton, A. W. 1944. Notes on the compressibility of clays. *Quaterly Journal of the Geological Society* 100: 119–135.

Skempton, A. W. 1954. The pore pressure coefficients A and B. *Geotechnique* 4: 143–147.

Skempton, A. W. and Bjerrum, L. 1957. A contribution to the settlement analysis of foundations on clay. *Geotechnique* 7: 168–178.

Skempton, A. W. and Northey, R. D. 1952. The sensitivity of clays. *Geotechnique* 3: 30–53.

Stroud, M. A. and Butler, F. G. 1975. The standard penetration test and the engineering properties of glacial materials. In: *Conference of the Midlands Geotechnical Society*. British Geotechnical Society, London.

Terzaghi, K. and Peck, R. B. 1967. *Soil Mechanics in Engineering Practice*. John Wiley, London, 729 pp.

Tomlinson, M. J. 1980. *Foundation Design and Construction*. Pitman, London, 793 pp.

US Navy. 1982. *Design Manual: Soil Mechanics, Foundations and Earth Structures*. Navy Facilities Engineering Command, Navfac, US Naval Publications and Forms Center.

Wroth, C. P. and Wood, D. M. 1978. The correlation of index properties with some basic engineering properties of soils. *Canadian Geotechnical Journal* 15: 137–145.

6

Shear Strength

Before considering soil shear strength in detail, it is perhaps worth considering why soil strength is measured in terms of its shear strength when other materials, such as metals and masonry materials like concrete and stone, are measured in terms of direct tensile or compressive strength.

Although it is convenient to think of strength in terms of a simple tensile strength for metals such as iron or steel, failure rarely occurs by simple snapping of a member, even in the case of a ductile metal such as steel, where failure during a tensile test is typically characterised by 'necking'; a pronounced elongation of part of a test section just prior to failure. This occurs as a result of metal atoms sliding over each other at about 45° to the direction of pull, so that what seems simplistically to be a failure in tension is in fact a failure by shearing along the lines of maximum shear stress.

For masonry materials such as concrete, failure during a crushing test is typically characterised by splitting of the test cube or cylinder at an angle of about 45° to its vertical axis, often resulting in a conical fracture surface indicating that failure is, again, really a shear strength phenomenon.

Within a mass of material such as soil which is subjected to compressive forces, no amount of pure compression will cause failure since the material is simply being squeezed into itself. In such a material, failure is brought

Soil Properties and their Correlations, Second Edition. Michael Carter and Stephen P. Bentley.
© 2016 John Wiley & Sons, Ltd. Published 2016 by John Wiley & Sons, Ltd.

about by differences in pressure in different directions, resulting in shear stresses, with failure occurring by shearing, as illustrated in Figure 6.1a.

6.1   Stresses Within a Material

6.1.1   The Mohr Diagram

The results of soil shear strength tests are normally plotted using the Mohr diagram, as illustrated in Figure 6.1d. In the case of triaxial tests, states of stress are represented by circles, often without much consideration of why it is that states of stress within a soil mass can be represented in this way. A better understanding of the variation of stresses throughout a soil mass, and the mode of failure within it, can be obtained by considering the stresses within the mass and justification for representing them as circles on a τ-σ Mohr plot.

Within a soil mass under pressure, the state of stress can always be represented by three principal stresses: σ_1 (maximum); σ_2 (intermediate); and σ_3 (minimum). Soil problems are normally linear, for instance long retaining walls, slopes or strip footings, in which no strain takes place at right angles to the cross-section. These 'plane strain' problems result in the pressure at right angles to the cross-section always being the intermediate principal stress, as indicated in the example of an idealised smooth cantilever retaining wall shown in Figure 6.2a. Since shear stresses depend on the difference between principal stresses, the critical combination of stresses will always be governed by the maximum and minimum principal stresses so only these need to be considered in stability analyses. In practical terms, this means that a retaining wall or slope, for instance, might be expected to fail by rotating or sliding forward but not by moving sideways – a fact that seems intuitively obvious. This greatly simplifies stability analyses.

In the triaxial test, σ_3 is the cell pressure acting on the side of the specimen while σ_1 is the sum of cell pressure and plunger pressure, acting on the ends of the specimen, as shown in Figure 6.2b. The intermediate principal stress σ_2 is the radial stress within the specimen, which is not measured.

6.1.2   Relationships of Stresses at a Point

Consider the normal stress σ on a plane at an angle θ to the σ_1 plane, as illustrated in Figure 6.1a. The component σ^* due to σ_1 (see Fig. 6.1b) is:

$$\sigma^* = \sigma_1 \cos\theta \div \frac{1}{\cos\theta} = \sigma_1 \cos^2\theta \tag{6.1}$$

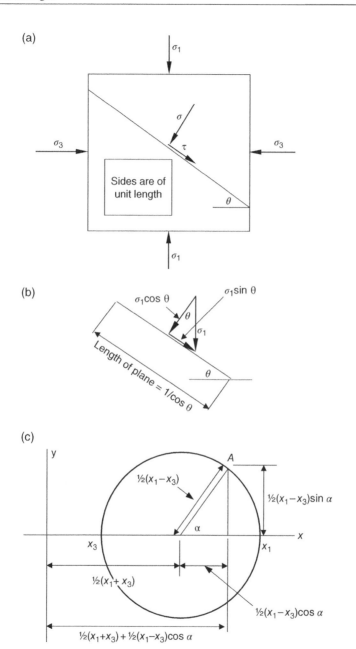

Figure 6.1 Stresses within a material and the Mohr circle construction: (a) stresses on a block of material; (b) resolution of stresses in relation to a sloping plane; (c) geometry of a circle in Cartesian co-ordinates; and (d) the Mohr circle representation of stresses.

(d)

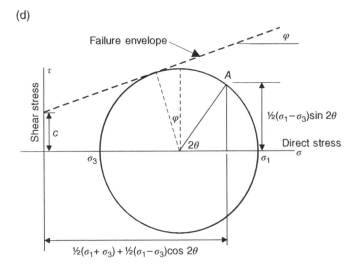

Figure 6.1 (*Continued*)

The component σ^{**} due to σ_3 is as for σ_1, but with the angle turned through $90°$:

$$\sigma^{**} = \sigma_3 \cos(90° - \theta) \div \frac{1}{\cos(90° - \theta)} = \sigma_3 \sin\theta \div \frac{1}{\sin\theta} = \sigma_3 \sin^2\theta. \quad (6.2)$$

Adding the stresses in Equations (6.1) and (6.3) gives

$$\sigma = \sigma^* + \sigma^{**} = \sigma_1 \cos^2\theta + \sigma_3 \sin^2\theta. \quad (6.3)$$

but $\cos^2\theta = \dfrac{1}{2}(1 + \cos 2\theta)$ and $\sin^2\theta = \dfrac{1}{2}(1 - \cos 2\theta)$.

Making these substitutions, Equation (6.3) becomes:

$$\sigma = \sigma_1 \frac{1}{2}(1 + \cos 2\theta) + \sigma_3 \frac{1}{2}(1 - \cos 2\theta) = \frac{1}{2}(\sigma_1 + \sigma_3) + \frac{1}{2}(\sigma_1 - \sigma_3)\cos 2\theta$$

$$(6.4)$$

Now consider the shear stress on the plane (see Fig 6.1a). The component τ^* due to σ_1 (see Fig. 6.1b) is:

$$\tau^* = \sigma_1 \sin\theta \div \frac{1}{\cos\theta} = \sigma_1 \sin\theta \cos\theta \quad (6.5)$$

(a)

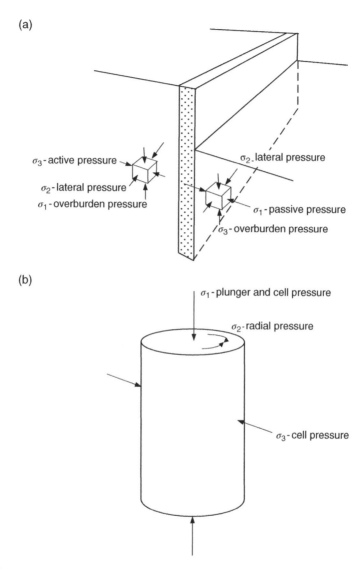

(b)

Figure 6.2 Examples of the principal stresses: (a) next to a smooth cantilever retaining wall; and (b) in a triaxial test specimen.

The component τ^{**} due to σ_3 is as for σ_1 but with the angle turned through $90°$, that is:

$$\tau^{**} = \sigma_3 \sin\left(90° - \theta\right) \div \frac{1}{\cos\left(90° - \theta\right)} = \sigma_3 \cos\theta \div \frac{1}{\sin\theta} = \sigma_3 \sin\theta \cos\theta.$$

(6.6)

Adding the stresses in Equations (6.5) and (6.6) to obtain τ, and remembering that the σ_3 component acts in the opposite direction to the σ_1 component, we obtain

$$\tau = \tau^* + \tau^{**} = \sigma_1 \sin\theta \cos\theta - \sigma_3 \sin\theta \cos\theta = \left(\sigma_1 - \sigma_3\right)\sin\theta \cos\theta \quad (6.7)$$

but $\sin\theta \cos\theta = \frac{1}{2}\sin 2\theta$. Making this substitution, Equation (6.7) becomes:

$$\tau = \frac{1}{2}\left(\sigma_1 - \sigma_3\right)\sin 2\theta \quad (6.8)$$

which will have a maximum value when $\sin 2\theta = 1$, that is, $2\theta = 90°$ and $\theta = 45°$.

Compare Equations (6.4) and (6.8) with the equations for a circle on the x-axis in Cartesian co-ordinates (Fig. 6.1c). For point A on the circle:

$$x = \frac{1}{2}\left(x_1 + x_3\right) + \frac{1}{2}\left(x_1 - x_3\right)\cos\alpha \quad (6.9)$$

and

$$y = \frac{1}{2}\left(x_1 - x_3\right)\sin\alpha. \quad (6.10)$$

It can be seen that Equations (6.4) and (6.8) are essentially the same as Equations (6.9) and (6.10), respectively, with the substitutions σ substituted for x, τ substituted for y and 2θ substituted for α. This means that we can represent the stress distribution on any plane within a body by a circle plotted on τ-σ co-ordinates, as shown in Figure 6.1d.

As well as the state of stress, the failure conditions can also be plotted, as illustrated, at the angle of shearing resistance φ, since φ represents the ratio in angular terms between shear and direct stresses. On examination of the figure, it can be seen that this shows the critical plane for failure occurs at an angle:

$$2\theta = 90° + \varphi, \quad (6.11)$$

that is,

$$\theta = 45° + \frac{1}{2}\varphi. \tag{6.12}$$

The Mohr circle construction can also be used for strains, plotting ½ shear strain against direct strain.

6.2 Shear Strength in Soils

The concept of soil shear strength was developed by Coulomb (1776) and then Mohr who suggested that the shear strength of soil could be characterised by a combination of a fixed cohesive component c, and a frictional component with a fixed angle of shearing resistance (or internal friction) φ. While this is somewhat simplistic in some ways, it provides a good working relationship for shear strength, and is the shear strength model that is mostly used. This leads to the commonly used Mohr-Coulomb failure criterion:

$$s = c + \sigma \tan \varphi \tag{6.13}$$

Where s is the shear stress at failure on any plane;
σ is the normal stress on that plane; and
c and φ are the shear strength parameters, cohesion and angle of shearing resistance.
This is shown graphically on the Mohr diagram given in Figure 6.3a.

A complication arises because the normal stresses within a soil are carried partly by the soil skeleton itself and partly by water within the soil voids. Considering only the stresses within the soil skeleton, Equation (6.13) is modified to:

$$s = c' + (\sigma - u)\tan \varphi' = c' + \sigma' \tan \varphi' \tag{6.14}$$

Where u is the pore-water pressure;
$\sigma' = (\sigma - u)$, the effective normal stress (on the soil skeleton); and
c' and φ' are the shear strength parameters related to effective stresses.

Thus when considering the shear strength of soils there is a choice: either the total, combined response of the soil and pore water can be considered (Equation (6.13)); or the specific response of the soil skeleton can be separated from the pore-water pressure by considering effective stresses (Equation (6.14)).

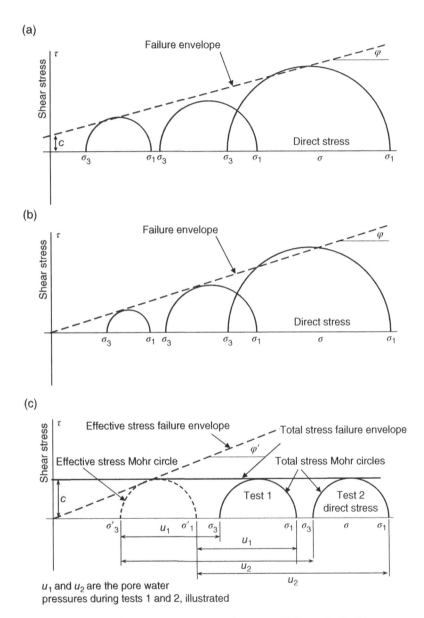

Figure 6.3   Mohr diagram representations of stress and failure criteria: (a) a generalised c-φ soil; (b) a purely frictional soil; and (c) a purely cohesive soil.

The effective stress approach gives a true measure of the response of the soil skeleton to the loads imposed on it. Perhaps the simplest case is that of a load applied to a saturated soil that is allowed to drain. If the rate of application of load is sufficiently slow, pore-water pressures will not build up and the effective stresses will equal the total stresses. For drained conditions, or in terms of effective stresses, it is found that the shear strength of soils is principally a frictional phenomenon with $c' = 0$, as illustrated in Figure 6.3b.

Hence, granular soils, which can drain rapidly so that excess pore-water pressures do not build up, tend to be entirely frictional so give the failure envelope shown in Figure 6.3b. Similarly, most clays, when tested in consolidated drained conditions or consolidated undrained conditions with pore-pressure measurement, also give a frictional failure envelope. Some clays, notably overconsolidated clays, also show a small drained or effective stress cohesion intercept, like that shown in Figure 6.3a, due to a built-in pre-stress (see Singh *et al.* 1973). Likewise, partially saturated clays in which the particles are drawn together by surface tension effects, exhibit some cohesion. However, many engineers do not like to take this into account in stability calculations as they believe that this cohesion, whatever its cause, is likely to degrade over time.

In contrast, the lower permeability of clay soils means that, when a clay is loaded, water cannot rapidly drain from the pores. In a saturated clay, the pore water therefore prevents the soil particles from squeezing together during loading. This means that any additional confining pressure within a saturated clay is taken by the pore water and not the soil skeleton. Since shear strength depends on the effective stresses, transmitted by inter-particle contacts, and these remain unchanged irrespective of the applied confining pressure, it follows that undrained shear strength will also be independent of confining pressure. Because of this, samples of saturated clay tested in a quick undrained triaxial test give (at least, in theory) Mohr's circles of constant diameter and an apparent cohesion value as shown in Figure 6.3c even though, in effective stress terms, the material is basically frictional. Thus, in a sense, the phenomenon of cohesion is an illusion brought about by the response of pore-water pressures to imposed loads. To underline this point, the term 'apparent cohesion' is often used. Partially saturated soils tested in undrained conditions will show a behaviour which is intermediate between that for drained conditions and for saturated undrained conditions, depending on the degree of saturation.

6.3   The Choice of Total or Effective Stress Analysis

When soil is loaded rapidly so that there is no time for movement of pore water to take place, its immediate response – the proportions of the resulting confining pressures that are carried by the soil skeleton and the pore water – is itself a property of the soil. This instantaneous response can, in fact, be quantified in terms of Skempton's (1954) pore-pressure parameters. This means that the total response of the soil to an applied load, including the pore pressures generated, can be simulated and measured in a laboratory test and there is no need to take account of the separate responses of the skeleton and the pore water. Only the total applied stresses need be considered in the analysis and only the corresponding total stress strength parameters need be measured when testing. Strictly speaking, this is not quite true because soil strength is usually measured in the triaxial test, in which axially symmetric stress conditions exist, whereas many soil problems approximate to plane strain conditions for which the soil response differs slightly but the errors involved are small enough to be ignored for practical purposes.

The equilibrium pore-water pressures that are eventually established are, unlike the immediate response, not a property of the soil but depend on the surrounding conditions. Long-term pore-water pressures cannot therefore be simulated in the laboratory and must be considered separately. Hence, effective stress analysis must be used where long-term stability is important. In testing, the response of the soil skeleton can be measured either by allowing drainage of the specimen so that no more pressures build up or by measuring the pore-water pressure within the specimen so that the stress on the soil skeleton can be determined. In either case, tests must be carried out slowly enough to allow complete dissipation or equalisation of excess pore-water pressures within the test specimen.

6.3.1   The Choice in Practice

Foundations impose both shear stresses and compressive stresses (confining pressures) on the underlying soil. The shear stresses must be carried by the soil skeleton but, in a saturated soil, the compressive stresses are initially carried largely by the resulting increase in pore-water pressures. This leaves the effective stresses little changed, which implies that the foundation loading is not accompanied by any increase in shear strength. As the excess pore pressures dissipate the soil consolidates and effective stresses increase, leading to an increase in shear strength. Therefore, for foundations, it is the short-term condition – the immediate response of the soil – that is most critical. This is the justification for the use of quick undrained shear strength tests and total stress analysis for foundation design.

With excavations, compressive stresses are reduced by removal of soil but shear stresses are imposed on the sides of the excavation owing to removal of lateral support. Initially, the reduction in compressive stresses is manifested within the soil mainly as a reduction in pore-water pressures with little change in effective stresses so that, as with foundations, soil shear strength remains little affected by the changed loading. Eventually, water flows into the soil that forms the excavation sides, restoring the pore-water pressures. This reduces the effective stresses, causes swelling and reduces shear strength. Thus, for excavations, long-term conditions are the most critical. Since long-term pore pressures depend on drainage conditions and cannot be simulated by soil tests, an effective stress analysis must be used so that pore-water pressures can be considered separately from stresses in the soil skeleton.

During embankment construction, additional layers of material impose a pressure on the lower part of the embankment. As with foundations, this tends to create increased pore-water pressures and, by the same argument, short-term conditions are an important consideration. This implies that total stress analysis and quick undrained shear strength tests are appropriate, and up to the 1960s it was not uncommon for embankments to be designed in this way. However, additional stresses can be created by the compaction process itself but, offsetting this, the material is unlikely to be saturated so that a significant proportion of the added pressures may be carried immediately by the soil skeleton. These complications make it impossible to simulate the total response of the soil in a test specimen and, to overcome this, effective stress analysis is now used. Also, until relatively recently it was often more economical to design embankments for long-term stability and to monitor pore-water pressures during construction, slowing down the rate of construction where necessary to keep them within safe limits. Alternatively, and more commonly in modern construction practice, soil reinforcement is used to deal with initial problems associated with a build-up of pore-water pressures. For either approach, effective stress analysis is needed.

With natural slopes, we are always dealing with conditions that have been in equilibrium for a long period of time, although seasonal variations will occur, and effective stress analysis is appropriate.

6.4   Peak, Residual and Constant-Volume Shear Strength

As described above, soils fail in shear once the shear stress along a critical plane within the mass of soil reaches a limiting value. As shearing takes place, grains along the plane of shear begin to roll over each other. Initially, the shearing forces must overcome both the frictional resistance between the grains and the interlock resistance. This gives rise to the peak shear strength. In most soils, this initial movement is accompanied by dilation; an increase in volume along the shear plane as the grains begin to ride over each other.

Once sufficient movement has occurred, the grains will roll over each other with no further dilation. This is the constant-volume condition which, because the grains along the shear plane are now in a loosened condition, gives rise to a slightly lower shear strength; the constant-volume shear strength. With further movement, resistance is further diminished, until the shear strength is eventually reduced to the residual value.

With normally consolidated clays there is often little dilation and the peak, constant volume and residual shear strengths may all be similar. However, with many soils, especially gravels, sands and overconsolidated clays, there may be a marked difference in strengths. As described in Chapter 1, it is usual to measure only peak shear strength in triaxial cells. Where residual strength is required, shear box testing will be needed or, for undrained shear strength, vane tests may be used; in both cases, large movements are applied to ensure that the shear strength has fully reduced to its minimum value.

The choice between these different shear strength values depends on the situation being analysed. For many situations, where no former movement has taken place and large movements are not anticipated during service, peak shear strength values are appropriate. This will normally apply in the cases of foundations, rigid retaining walls and new or stable slopes. However, for slopes that have already failed or are in an area with a history of landslip or surface creep, residual shear strength is more appropriate. For reinforced earth structures, using geotextiles or geogrids, where significant but not excessive movement is anticipated as the flexible reinforcement takes up the load, constant-volume shear strength may be specified for the design procedure. A problem arises here in that constant-volume shear strength is not normally measured in testing, so it is often estimated by applying a reduction of typically 20% to the measured peak strength.

6.5 Undrained Shear Strength of Clays

As described previously, shear strength is obtained from the Mohr-Coulomb failure criterion, Equation (6.13). However, for most saturated clays tested under quick undrained conditions the angle of shearing resistance is zero. This means that the shear strength of the clay is a fixed value and is equal to the apparent cohesion.

Typical values for the shear strengths of compacted clays are given in Table 6.1. Values refer to soils compacted to the maximum dry density obtained in the standard compaction test (AASHTO 1982, T99, 5.5 lb rammer method; or BS 1990, 1377, Test 12, 2.5 kg rammer method).

6.5.1 Consistency and Remoulded Shear Strength

As discussed in Chapter 2, the liquid and plastic limits are moisture contents at which soil has specific values of undrained shear strength. It therefore follows that, for a remoulded soil, the shear strength depends on the value of the natural moisture content in relation to the liquid and plastic limit values. This can be conveniently expressed by using the concept of liquidity index (LI) or consistency index (CI), defined as:

$$LI = \frac{m - PL}{LL - PL} = \frac{m - PL}{PI} \qquad (6.15)$$

and

$$CI = \frac{LL - m}{LL - PL} = \frac{LL - m}{PI} \qquad (6.16)$$

Table 6.1 Typical values of the undrained shear strength of compacted soils.

Soil description	Class*	Undrained shear strength (kPa)	
		As compacted	Saturated
Silts, sands, sand-silt mix	SM	50	20
Clayey sands, sandy clay mix	SC	75	10
Silts and clayey silts	ML	65	10
Clays of low plasticity	CL	85	15
Clayey silts, elastic silts	MH	70	20
Clays of high plasticity	CH	100	10

* Unified classification system.

where LL and PL are the liquid and plastic limits, respectively;
        PI is the plasticity index;
and m is the soil moisture content.

Basically, both expressions define the moisture content in relation to the values over which the soil is plastic; liquidity index defines it in terms of how far it is above the plastic limit, and the consistency index defines it in terms of how far it is below the liquid limit. The relationship between the two is therefore:

$$LI = 1 - CI. \qquad (6.17)$$

Curves relating remoulded undrained shear strength to liquidity index have been established by Skempton and Northey (1952). These are given in Figure 6.4. It can be seen from this that, when shear strength is plotted to a logarithmic scale, the relationship is nearly linear. This gives rise to a simple relationship between remoulded shear strength s and liquidity index or consistency index:

$$s = n10^{2(1-LI)} \quad \text{or} \quad s = n10^{2CI} \qquad (6.18)$$

Where s is the undrained remoulded shear strength and
        n is a constant. If s is in kPa then,
            if the shear strength at the liquid and plastic limits is 1 kPa and
            100 kPa, respectively, then $n = 1$;
            if the shear strength at the liquid and plastic limits is 2 kPa and
            200 kPa, respectively, then $n = 2$.

This relationship is plotted on Figure 6.4 for both $n = 1$ and $n = 2$, from which it can be seen that $n = 1$ gives a better fit to the measured strengths but can still overestimate shear strength by a factor of 2 or 3.

6.5.2   Consistency and Undisturbed Shear Strength

6.5.2.1   Estimates Based on Liquidity/Consistency Index and Sensitivity

The shear strength of undisturbed clays depends on the consolidation history of the clay as well as the fabric characteristics. Clay in its natural state can therefore be expected to have a different strength from that in its

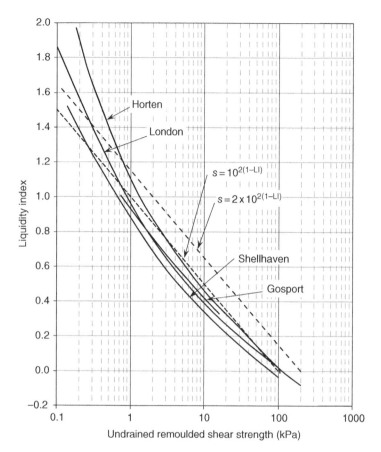

Figure 6.4 Correlations between remoulded shear strength and liquidity index.
Reproduced with permission of Skempton and Northey (1952).

remoulded, compacted state. This makes shear strength prediction of undisturbed soils from consistency limits even less reliable.

The ratio of undisturbed shear strength to remoulded shear strength is known as the sensitivity. It is most marked in soft, lightly consolidated clays which have an open structure and a high moisture content. It therefore seems reasonable that sensitivity should be related to liquidity index, and this has been found to be the case by a number of researchers, whose findings are given and discussed by Holtz and Kovacs (1981) and Holtz *et al.* (2011). Much of this data is for the highly sensitive clays of Canada and

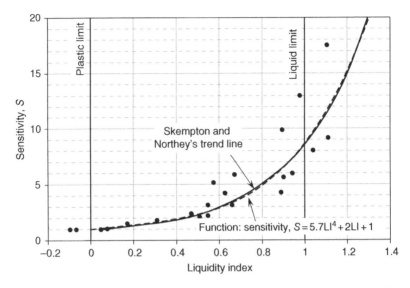

Figure 6.5 Correlation between sensitivity and liquidity index. Reproduced with permission of Skempton and Northey (1952).

Scandinavia but the work of Skempton and Northey (1952) relates mainly to clays of relatively moderate sensitivity with natural moisture contents below the liquid limit. Their findings for liquidity index values up to 1.3 are given in Figure 6.5, which also shows a plot of the function:

$$\text{Sensitivity}, S = 5.7\text{LI}^4 + 2\text{LI} + 1. \qquad (6.19)$$

This shows a good fit to the data points for soils with a liquidity index of up to 1.3.

6.5.2.2   Estimates Based on Consistency Descriptions

Most descriptions of clay soils begin with a consistency description such as 'soft', 'firm' or 'stiff'. Originally, such terms were related the soil's resistance to being moulded in the hands as shown in Table 6.2. Since moulding essentially means shearing, it follows that consistency defined in this way is a crude measure of shear strength, the simple moulding tests listed in Table 6.2 being basically shear strength tests. With the desire for more uniformity in soil descriptions, the consistency descriptions became associated

Table 6.2   Values of undrained shear strength based on consistency descriptions.

Consistency description	Characteristics	Strength ranges[a] (kPa)	EC7 definitions of consistency and estimated shear strength ranges[b]		
			Liquidity index	Consistency index	Strength ranges (kPa)
Very soft	Exudes between fingers when squeezed	<20	>0.75	<0.25	<50
Soft	Moulded by light finger pressure	20–40	0.5–0.75	0.25–0.5	25–70
Firm	Moulded by strong finger pressure	40–75	0.25–0.5	0.5–0.75	35–100
Stiff	Can be indented by thumb	75–150	0.0–0.25	0.75–1.0	50–100
Very stiff	Can be indented by thumb nail	150–300	<0.0	>1.0	70–140
Hard		>300			

[a] Undrained shear strength as defined by BS5930 (BS 1981).
[b] Based on liquidity/consistency index values.

with specific undrained shear strengths that roughly corresponded to the values obtained by these moulding tests. The corresponding shear strength values shown in column 3 of the table were adopted in British Standard BS5930 as the defining strengths for consistency descriptions. This meant that, from the consistency term used in the soil description, an engineer could have an estimate of the range of undrained shear strength of a soil.

However, Eurocode EC7 (2004) now defines consistency in terms of consistency index or liquidity index as shown in Table 6.2. An indication of the range of shear strength that could occur for a given range of liquidity index can be obtained by taking the shear strength based on liquidity index (for $n = 1$ and $n = 2$ to give lower and upper bound limits), with a modification to allow for sensitivity, which is itself related to liquidity index. This approach was used by the authors to give the strength ranges for undisturbed soils based on consistency as defined in EC7. The resulting values are shown in the last column of Table 6.2 from which it can be seen that consistency descriptions using EC7 definitions give wide ranges of shear strength for undisturbed soils. This concurs with the authors' own experience of the use of liquidity index to estimate shear strength. Direct measurement of undrained shear strength, even using simple equipment such as a hand vane or hand penetrometer, gives far more reliable results; given the ease and speed of such testing, there seems little merit in using liquidity index to estimate undisturbed shear strength.

6.5.2.3 Estimates Based on Overburden Pressure

It is found that for most normally consolidated clays, undrained shear strength is proportional to effective overburden pressure. This is to be expected when it is remembered that, in terms of effective stress, shear strength is basically a frictional phenomenon and depends on confining pressure. If the constant of proportionality between shear strength and effective overburden pressure is known then shear strength can be inferred from effective overburden pressure, that is, from depth. This problem has been investigated by a number of researchers, with a view to establishing a correlation between the shear-strength/overburden-pressure ratio and some soil classification parameter, typically the plasticity index. Such a correlation would be of great practical value, since it would enable the undrained shear strength c_u to be estimated from a simple classification test.

Historically, for normally consolidated clays, much use has been made of the relationship of Skempton and Bjerrum (1957):

$$c_u = (0.11 + 0.0037\,PI)\sigma'_v \qquad (6.20)$$

Where c_u is the undrained shear strength,
PI is the plasticity index and
σ'_v is the effective overburden pressure.

At first sight it is not evident that c_u/σ'_v should be related to plasticity index. However, the values of both c_u and PI can be expected to depend on the shape, size and mineral composition of the clay particles, so the two properties can be expected to be related in some manner. Figure 6.6 shows Skempton's relationship along with the results of other researchers. As can be seen, their findings vary and should be used with caution. However, such correlations, particularly that of Skempton, can be used for preliminary estimates on normally consolidated clays. For overconsolidated clays, Kenney (1959) stated that the relationship is influenced mainly by the stress history and is essentially independent of plasticity index. A correlation between the

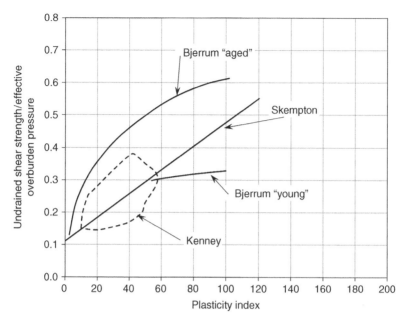

Figure 6.6   Relationship between the ratio of undrained shear strength to effective overburden pressure and plasticity index for normally consolidated clays. Adapted from Holtz and Kovacs (1981).

shear-strength/overburden-pressure ratio and liquidity index for Norwegian quick clays was presented by Bjerrum and Simons (1960), as indicated in Figure 6.7. Again, results show so much scatter, especially within the liquidity index range for most soils of 0 to 1, that the usefulness of the correlation is open to question.

A further uncertainty arises with overconsolidated clays, whose strength is related to the previous consolidation history rather than the current consolidation pressure. This is where Figure 6.7 should in theory be useful, but in practice the consolidation history of a deposit may be difficult to determine. Although the literature contains much debate concerning s_u/σ'_v ratios and overconsolidation ratios (Ladd *et al.* 1977; Wroth 1984), in practical terms it is more straightforward to measure the undrained shear strength of overconsolidated clays than to try and predict it from correlations with other values. Therefore, no correlations with overconsolidation ratios have been included here.

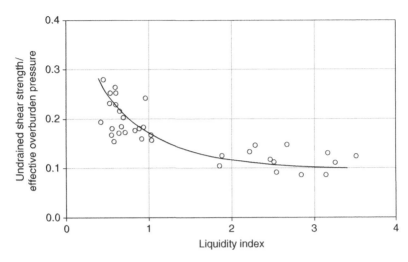

Figure 6.7 Correlations between shear strength and liquidity index. Adapted from Bjerrum and Simons (1960). Reproduced with permission of American Society of Civil Engineers.

Besides the influence of geological history on undrained shear strength, the stress history during test also affects results. Shear strengths obtained by unconfined compression testing or triaxial testing can therefore be expected to differ from those obtained by shear vane (Wroth 1984). The relative values of the shear strengths have been examined by a number of researchers, and the ratio of 'true' undrained shear strength (based on the back-analysis of embankment failures) to shear vane values seems to depend on the plasticity index, as indicated by Figure 6.8.

6.5.3   Estimates Using the Standard Penetration Test

Attempts have been made to correlate the unconfined compressive strength or the undrained shear strength of clays with the results of standard penetration tests, with varying degrees of success. Some suggested relationships are given in Figure 6.9.

Probably one of the most commonly used relationships is that given by Stroud and Butler (1975), in which the relationship between SPT N-values and shear strength is not fixed but is modified by the plasticity index value, as shown in Figure 6.10. Although this correlation is better than many, it can be seen that there is a significant scatter of results with c_u/N values ranging from about 10% higher to about 20% lower than those predicted by the correlation.

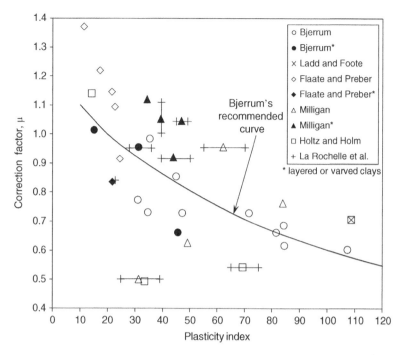

Figure 6.8 Correction factor for field vane test results depending on plasticity index, based on back-analysis of embankment failures. Adapted from Bjerrum and Simons (1960). Reproduced with permission of American Society of Civil Engineers.

Although the original paper presents the relationship graphically, the authors have established the following mathematical relationship that closely follows the original graphical curve, and may be used in spreadsheet calculations:

$$\frac{C_u}{N} = \frac{8910}{PI^3} + 4.36 \tag{6.21}$$

Although the Stroud and Butler correlation is widely used, it is often forgotten that it was developed for use with overconsolidated soils and that all the soils shown on the graph are overconsolidated deposits. For normally consolidated soils it should be used, if at all, with caution. The misuse of the relationship in this way is discussed by Reid and Taylor (2010). Ideally, it should be checked against consolidation test results for the specific site and soil type, then used to extend the quantity of results from SPT N-values once the correlation has been validated.

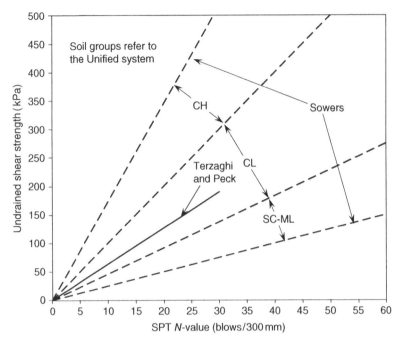

Figure 6.9 Approximate correlations between undrained shear strength and SPT *N*-values. Adapted from Terzaghi and Peck (1967) and Sowers (1979).

6.5.4 Estimates Using Dynamic Cone Tests

Like standard penetration testing, dynamic probe testing is more suited to granular soils than clays and there is no simple relationship between probe count and undrained shear strength. Relationships were, however, developed by Butcher *et al.* (1995) between various types of clay, related to sensitivity and 'dynamic cone resistance' q_d, defined by:

$$q_d = \frac{M_h}{M_h + M_a} r_d \qquad (6.22)$$

Where M_h is the mass of the hammer (50 kg),
$\quad M_a$ is the combined mass of the anvil, guiding rods and extension rod (18 kg + 6 kg/m depth) and
$\quad r_d$ is the 'unit point resistance', defined by:

$$r_d = \frac{M_h \, g \, H}{A \, e} \qquad (6.23)$$

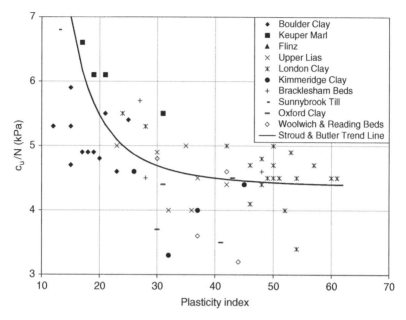

Figure 6.10   Correlation between undrained shear strength, SPT N-value and plasticity index for overconsolidated clays. Adapted from Stroud and Butler (1975). Reproduced with permission of Midland Geotechnical Society.

where H is the fall of the hammer (0.5 m),

  A is the cone tip area (0.015 m^2),

  e is the average cone penetration (m) per blow and

  g is the acceleration due to gravity (9.81 m/s^2).

The figures given in brackets are the values for a standard Heavy Dynamic cone.

For a clay of known sensitivity S (the ratio of undisturbed shear strength to remoulded shear strength), the shear strength of the clay is estimated from the basic relationship:

$$c_u = \frac{0.045 q_d}{S} + 10. \qquad (6.24)$$

Where sensitivity is not known, c_u may be estimated according to clay type using the relationships:

$$c_u = \frac{q_d}{22} \qquad \text{for a stiff clay}$$

$$\text{or } c_u = \frac{q_d}{170} + 20 \qquad \text{for a soft clay.}$$

The relationships for a stiff clay and a soft clay correspond approximately to values obtained from the basic relationship for sensitivity values of 1.2 and 4, respectively.

In addition, the method gives a relationship between the dynamic cone blow count and an equivalent SPT N-value:

$$N = 8N_p - 6 \qquad\qquad (6.25)$$

Where N is the SPT N-value (blows/300 mm) and
        N_p is the probe blow count (blows/100 mm).

It can be that the shear strength value obtained from the proposed relationships depends on a prior knowledge of the sensitivity or clay stiffness; that is, on its shear strength. A value of a low sensitivity or a presumption that the clay is stiff (has a high shear strength) will result in a high predicted shear strength, whereas a high sensitivity or a presumption that the clay is soft (has a low shear strength) will result in a low predicted shear strength, making the method something of a self-fulfilling prophesy. It is therefore only suitable where dynamic probes are used between boreholes as a check on variability, and the clay strength has been determined by direct measurement from samples taken at the borehole locations. This will severely limit its use.

6.5.5   Estimates Using Static Cone Tests

A preliminary estimate of shear strength c_u can be obtained from the cone resistance q_c, using the relationship:

$$c_u = \frac{q_c}{N_k} \qquad\qquad (6.26)$$

where N_k is usually taken as 17 or 18 for normally consolidated clays and 20 for overconsolidated clays.

6.6   Drained and Effective Shear Strength of Clays

As discussed previously in Section 6.3 it is often important to carry out stability calculations in terms of effective stresses. This is particularly true of slope stability calculations. The soil strength parameters used in these calculations are obtained from either drained shear box or triaxial tests

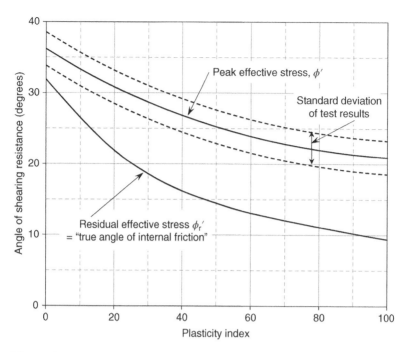

Figure 6.11 Correlations between effective angle of shearing resistance and plasticity index. Adapted from Gibson (1953). Reproduced with permission of Geotechnik Schweiz.

(giving c_d and φ_d) or from consolidated undrained triaxial tests with pore pressure measurement (giving c' and φ'). In theory there should be minor differences between the two sets of values but in practical terms, for normal laboratory testing, the differences are normally of no consequence.

A relationship between drained shear strength and plasticity index for remoulded clays has been established by Gibson (1953), as indicated in Figure 6.11. Also shown is a relationship between residual shear strength, or true angle of internal friction, and plasticity index. The existence of these relationships arises because both plasticity index and shear strength reflect the clay mineral composition of the soil; as the clay mineral content increases, plasticity index increases and shear strength decreases. As described previously, the strength of clays in effective stress terms is basically frictional so $c' = 0$. This is certainly the case with remoulded saturated clays but partially saturated clays, where meniscus effects draw the particles together

Table 6.3 Typical values of the effective angle of shearing resistance of compacted clays.

Soil description	Class*	φ' (°)
Silty clays, sand-silt mix	SM	34
Clayey sands, sand-clay mix	SC	31
Silts and clayey silts	ML	32
Clays of low plasticity	CL	28
Clayey silts, elastic silts	MH	25
Clays of high plasticity	CH	19

* Unified classification system.

to produce inter-particle stresses, may appear to have a small cohesion value, though this itself is a frictional phenomenon.

For use with spreadsheets, the correlations shown in Figure 6.11 may be calculated by using the following polynomial relationships:

for peak strength, $\varphi' = 0.0058\,\mathrm{PI}^{1.73} - 0.32\,\mathrm{PI} + 36.2$ (6.27)

for residual strength, $\varphi'_r = 0.084\,\mathrm{PI}^{1.4} - 0.75\,\mathrm{PI} + 31.9$ (6.28)

The polynomial for peak strength gives an almost-perfect fit to the original curves, with agreement to within 1%, while the polynomial for residual strength gives a good approximation, to within 5% of the original curves.

Typical values of the angle of effective shearing resistance φ' for compacted clays are given in Table 6.3. Values are for soils compacted to the maximum dry density according to the standard compaction test (AASHTO 1982, T99, 5.5 lb rammer method; or BS 1990, 1377, test 12, 2.5 kg rammer method).

6.7   Shear Strength of Granular Soils

Because of their high permeability, pore-water pressures do not build up when granular soils are subjected to shearing forces as they do with clays. The complication of total and effective stresses is therefore avoided and the phenomenon of apparent cohesion, or undrained shear strength, does not occur. Consequently, the shear strength of granular soils is defined exclusively in terms of the frictional resistance between the grains, as measured by the angle of shearing resistance.

Table 6.4 Typical values of the angle of shearing resistance of cohesionless soils.

Soil description	φ (°)	
	Loose	Dense
Uniform sand, round grains	27	34
Well-graded sand, angular grains	33	45
Sandy gravel	35	50
Silty sand	27–33	30–34
Inorganic silt	27–30	30–35

Table 6.5 Typical values of the angle of shearing resistance of compacted sands and gravels.

Soil description	Class*	φ (°)
Well graded sand-gravel mixtures	GW	>38
Poorly graded sand-gravel mixtures	GP	>37
Silty gravels, poorly graded gravel-sand-silt	GM	>34
Clayey gravels, poorly graded gravel-sand-clay	GC	>31
Well-graded clean sand, gravelly sand	SW	38
Poorly graded clean sand, gravelly sand	SP	37

* Unified classification system.

Typical values of the angle of shearing resistance for sands and gravels in loose and dense conditions are given in Table 6.4. Typical values for soils compacted to the maximum dry density according to the standard compaction test (AASHTO T99, 5.5 lb rammer method; or BS 1377:1975 test 12, 2.5 kg rammer method) are given in Table 6.5.

Both dry density and angle of shearing resistance can be estimated from a knowledge of relative density (obtained from SPT N-values) and material type using relationships given by the US Navy (1982) as shown in Figure 6.12. The material types indicated in the figure relate to the Unified classification system. Peck et al. (1974) give a correlation of angle of shearing resistance with standard penetration test values, as shown in Figure 6.13. The correlation between SPT N-values and relative density is also shown, enabling a comparison to be made with the US Navy estimates of the angle of shearing resistance.

Examination of Figures 6.12 and 6.13 shows reasonable agreement between the two correlations. However, considerable variation can exist within each soil type, as indicated by Figure 6.14 which shows plots of the

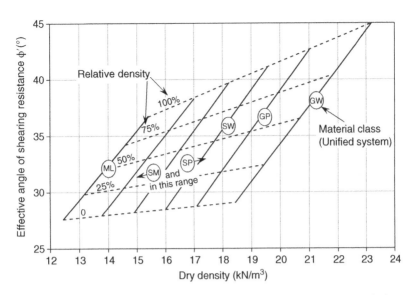

Figure 6.12 Typical values of density and angle of shearing resistance for cohesionless soils. Adapted from US Naval Publications and Forms Center, US Navy (1982).

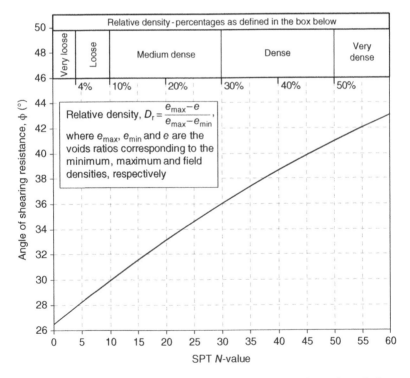

Figure 6.13 Estimation of the angle of shearing resistance of granular soils from standard penetration test results. Adapted from Peck *et al.* (1974). Reproduced with permission of John Wiley & Sons.

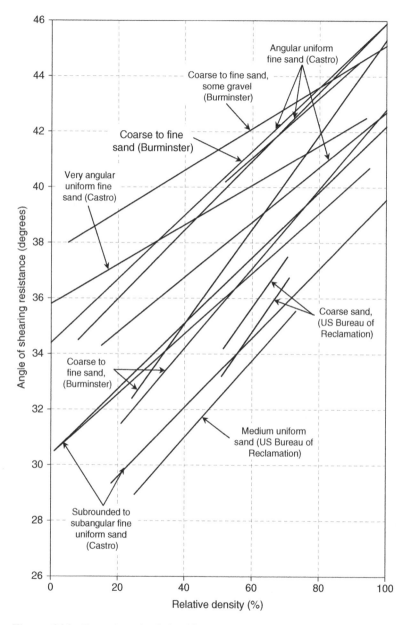

Figure 6.14 Examples of relationships between relative density and angle of shearing resistance. Adapted from US Naval Publications and Forms Center, US Navy (1982).

angle of shearing resistance against relative density of a number of sands, and provides a warning against putting too much reliance on estimates of the angle of shearing resistance based on relative density.

References

AASHTO. 1982. *Standard specifications for transportation materials and methods of testing and sampling.* American Association of State Highway and Transportation Officials. Washington, DC, USA.

Bjerrum, L. and Simons, N. E. 1960. Comparison of shear strength characteristics of normally consolidated clays. *Proceedings of ASCE Research Conference on the Shear Strength of Cohesive Soils,* Boulder: 711–726.

BS. 1981. *Code of practice for site investigation.* British Standards Institution, BS 5930.

BS. 1990. *Methods of test for soils for civil engineering purposes.* British Standards Institution, BS 1377.

Burmister, D.M. 1954. Principles of permeability testing of soils, symposium on permeability of soils. *ASTM Special Technical Publication* 163: 3–26.

Butcher, A. P., McElmeel, K. and Powell, J. J. M. 1995. Dynamic probing and its use in clay soils. *Advances in Site Investigation Practice.* Thomas Telford, London.

Castro, G. 1969. *Liquefaction of Sands.* Harvard Soil Mechanics Series, No. 81, 1 12 pp.

Coulomb, C.A. 1776. Essai sur une application des regles de maximis et minimis à quelques problèmes de statique, relatifs à l'architecture. *Memoires de l'Academie Royale des Sciences, Paris,* 3, p38.

Eurocode 7. 2004. *Geotechnical Design.* BS EN 1997-1: 2004.

Gibson, R. E. 1953. Experimental determination of the true cohesion and true angle of internal friction in clays. *Proceedings of 3rd International Conference on Soil Mechanics and Foundation Engineering,* Zurich, 126–130.

Holtz, R. D. and Kovacs, W. D. 1981. *An Introduction to Geotechnical Engineering.* Prentice-Hall, New Jersey, 733 pp.

Holtz, R. D., Kovacs, W. D. and Sheahan, T. C. 2011. *An Introduction to Geotechnical Engineering.* Pearson, New Jersey, USA.

Kenney, T. C. 1959. Discussion of geotechnical properties of glacial lake clays. *Proceedings ASCE Journal of Soil Mechanics and Foundation Division* 85: 67–79.

Ladd, C. C., Foote, R., Ishihara, K., Schlosser, F. and Poulos, H. G. 1977. Stress-deformation and strength characteristics. *Proceedings of 9th International Conference on Soil Mechanics and Foundation Engineering,* Tokyo, 2: 421–494.

Peck, R. B., Hanson, W. E. and Thornburn, T. H. 1974. *Foundation Engineering.* John Wiley, London, 514 pp.

Reid, A. and Taylor, J. 2010. The misuse of SPTs in fine soils and the implications of Eurocode 7. *Ground Engineering* July 2010, 28–31.

Singh, R., Henkel, D. J. and Sangay, D. A. 1973. Shear and K0 swelling of over-consolidated clay. *Proceedings of 8th International Conference on Soil Mechanics and Foundation Engineering,* 367–376.

Skempton, A. W. 1954. The pore pressure coefficients A and B. *Geotechnique* 4: 143–147.

Skempton, A. W. and Bjerrum, L. 1957. A contribution to the settlement analysis of foundations on clay. *Geotechnique* 7: 168–178.

Skempton, A. W. and Northey, R. D. 1952. The sensitivity of clays. *Geotechnique* 3: 30–53.

Sowers, G. F. 1979. *Introductory Soil Mechanics and Foundations.* Macmillan, New York.

Stroud, M. A. and Butler, F. G. 1975. The standard penetration test and the engineering properties of glacial materials. *Conference of the Midlands Geotechnical Society.*

Terzaghi, K. and Peck, R. B. 1967. *Soil Mechanics in Engineering Practice.* John Wiley, London, 729 pp.

US Bureau of Reclamation. 1974. *Earth Manual.* Denver, 810 pp.

US Navy. 1982. *Design Manual: Soil Mechanics, Foundations and Earth Structures.* Navy Facilities Engineering Command, Navfac, US Naval Publications and Forms Center.

Wroth, C. P. 1984. The interpretation of in situ soil tests. *Geotechnique* 34: 449–489.

7

California Bearing Ratio

The California Bearing Ratio test, or CBR as it is usually known, was originally developed by the California Division of Highways in the 1930s as part of a study of pavement failures. Its purpose was to provide an assessment of the relative stability of fine crushed rock base materials. Later its use was extended to subgrades. It was subsequently used for pavement design throughout the world and, despite numerous criticisms and a fall in popularity over recent years, still forms the basis of a number of current pavement thickness design methods. Ironically, it was used for pavement design in California for only a few years, and was superseded by the Hveem Stabilometer.

The CBR test is used exclusively in conjunction with pavement design methods. As described in Chapter 1, there are a number of variations in the test method, and results depend critically on both the method of specimen preparation and the test procedure, which follow the assumptions made in the design method. For instance, pavement design methods that follow US practice tend to assume that the specimen will always be soaked before testing, regardless of actual site conditions, with the design method itself incorporating a climatic factor; whereas UK practice has been to test the specimen at the worst long-term moisture content conditions anticipated on site, with no allowance for climate in the design method itself. Some differences that affect results are indicated in Table 7.1.

Soil Properties and their Correlations, Second Edition. Michael Carter and Stephen P. Bentley.
© 2016 John Wiley & Sons, Ltd. Published 2016 by John Wiley & Sons, Ltd.

Table 7.1   Some variations that can affect CBR test results.

Variation	Comment
Density	The CBR is usually quoted for the assumed density of the soil in place. This will typically be 90%, 95% or 100% dry density, as specified in either a standard (2.5 kg rammer) or heavy (4.5 kg rammer) compaction test.
Moisture content	The aim is to test the specimen under the worst likely conditions that will occur within the subgrade. In practice, soil is usually compacted at either optimum moisture content as specified in a compaction test or anticipated worst field moisture content. It is then either tested immediately or soaked for 4 days before testing, depending on the pavement thickness design method used.
Surcharge weights	Surcharge weights are placed on the specimen before testing to simulate the weight of pavement materials overlying the subgrade. In practice, 2 or 3 weights are usually used but this can vary. The effect of the surcharge weights is more marked with granular soils.
Testing top and bottom faces	It is usual US practice to test the bottom of the specimen since soaking significantly softens the top face. In Britain, where soaking is not usually specified, both top and bottom faces are tested and the average taken. Since the top face usually gives a lower CBR value than the bottom face, this variation can significantly affect results.
Method of compaction	The AASHTO specification stipulates the use of dynamic compaction (using a rammer) for sample preparation, but the BS specification allows the use of static compaction (using a load frame) or dynamic compaction (using either a rammer or a vibrating hammer).
In situ values	If tests are carried out on completed construction, the lack of the confining influence of the mould and drying out of the surface can affect results.

7.1   Correlations with Soil Classification Tests

In view of the fact that early pavement design methods were based on soil classification tests rather than CBR values, it seems a reasonable assumption that CBR values are related to soil classification in some way. However, CBR values depend not only on soil type but also on density, moisture content and, to some extent, method of preparation. These factors must therefore be taken into account when considering correlations between CBR and soil classification tests.

Table 7.2   Estimated CBR values for British soils compacted at the natural moisture content. Adapted from TRRL (1970).

Type of soil	Plasticity index	CBR (%) Depth of water table below formation level	
		More than 600 mm	Less than 600 mm
Heavy clay	70	2	1
	60	2	1.5
	50	2.5	2
	40	3	2
Silty clay	30	5	3
	20	6	4
	10	7	5
Silt	–	2	1
Sand (poorly graded)	Non-plastic	20	10
Sand (well graded)	Non-plastic	40	15
Well-graded sandy gravel	Non-plastic	60	20

A number of attempts have been made to correlate CBR with soil plasticity. A correlation between plasticity index and CBR, for design purposes, was given by the Transport and Road Research Laboratory (1970) as indicated in Table 7.2. This is based on wide experience of subgrade soil but is limited to British soils compacted at natural moisture content according to the Ministry of Transport (1969) specification. The precise density and moisture content conditions corresponding to the given CBR values are therefore not specified, limiting the usefulness of the table outside Britain.

The values used by the Transport and Road Research Laboratory owe much to the work of Black (1962), who obtained correlations between CBR and plasticity index for various values of consistency index (defined in Chapter 6) as shown in Figure 7.1. The values obtained from Figure 7.1 refer to saturated soils. For unsaturated soils, the CBR can be estimated by applying a correction to the saturated value using Figure 7.2.

Morin and Todor (1977) report attempts to correlate soaked CBR values with a modified plasticity index (sometimes called plasticity modulus) for tropical African and South American soils compacted at optimum moisture content and maximum dry density, where the modified plasticity index PM is defined as:

$$PM = PI \times \text{percent passing } 425\,\mu\text{m sieve.} \qquad (7.1)$$

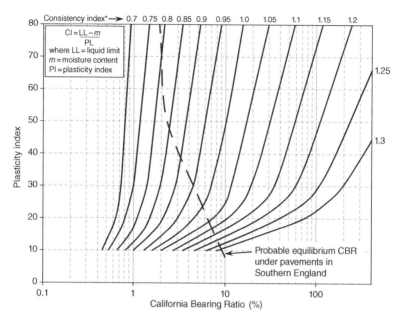

Figure 7.1 Relationship between CBR and plasticity index for various consistency index values. Adapted from Black (1962). Reproduced with permission of ICE.

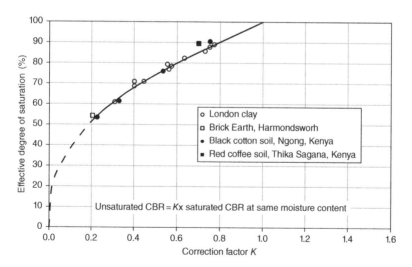

Figure 7.2 Correction of CBR values for partial saturation. Adapted from Black (1962). Reproduced with permission of ICE.

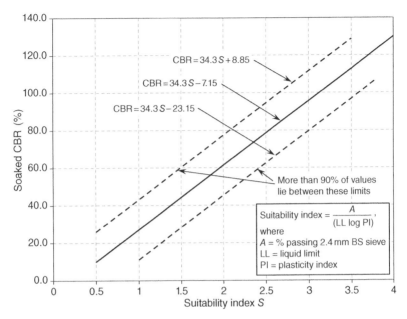

Figure 7.3 Relationship between suitability index and soaked CBR values. Adapted from de Graft-Johnson and Bhatia (1969).

They concluded that no well-defined relationship existed. However, de Graft-Johnson and Bhatia (1969) obtained a correlation of CBR with plasticity and grading for lateritic gravels for road pavements using the concept of suitability index, defined by:

$$\text{Suitability index} = \frac{A}{LL log \text{PI}} \qquad (7.2)$$

where A is the percentage passing a 2.4 mm BS sieve and LL and PI are liquid limit and plasticity index, respectively.

Their findings are given in Figure 7.3. Note, however, that the CBR values are for samples compacted to maximum dry density at optimum moisture content according to the Ghana standard of compaction. This specified the use of a standard CBR mould and a 4.5 kg rammer with a 450 mm drop to compact soil in 5 layers using 25 blows per layer. Samples are tested after a 4-day soak.

Further work on lateritic gravels (de Graft-Johnson *et al.* 1972) led to the establishment of a relationship between CBR and the ratio of maximum dry density to plasticity index, as shown in Figure 7.4.

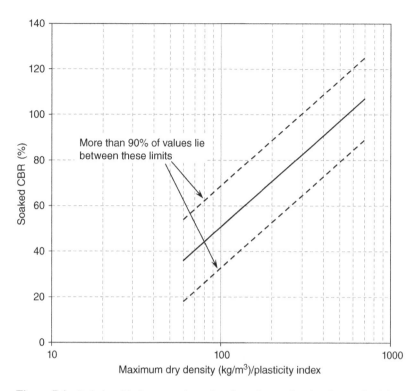

Figure 7.4 Relationship between the ratio of maximum dry density to plasticity index and CBR for laterite–quartz gravels. Adapted from de Graft-Johnson *et al.* (1972).

Based on tests of 48 Indian fine-grained soils, Agarwal and Ghanekar (1970) found no significant correlation between CBR and either liquid limit, plastic limit or plasticity index. However, they did obtain better correlations when optimum moisture content was taken into account. The best-fit relationship was for CBR with optimum moisture content (OMC) and liquid limit (LL):

$$CBR = 21 - 16\log(OMC) + 0.07LL. \tag{7.3}$$

The soils tested all had CBR values of less than 9 and the standard deviation obtained was 1.8. They therefore suggest that the correlation is only of sufficient accuracy for preliminary identification of materials. They further suggest that such correlation may be of more use if derived for specific geological regions.

More recent work on Indian soils by Shirur and Hiremath (2014) also gave no correlation between CBR and liquid or plastic limits but did find correlations with plasticity index, maximum dry density and optimum moisture content as indicated in Figure 7.5. Testing was carried out according to Indian standards, which require a 4-day soak for CBR. Using multiple linear regression analyses, Shirur and Hiremath also found that good predictions of CBR values could be obtained from the relationships:

$$CBR = 4.64MDD - 1.57OMC - 4.84; \qquad (7.4)$$

$$CBR = 2.8MDD - 0.069PI - 3.24; \quad and \qquad (7.5)$$

$$CBR = 6.54 - 0.077OMC - 0.1PI. \qquad (7.6)$$

The upper graph in Figure 7.5 would be useful for preliminary assessment of CBR values where only plasticity test data is available but, where compaction testing is carried out, it would be better to specify a CBR mould for the compaction tests and carry out CBR tests on the compaction test specimens rather than relying on correlations.

7.2   Correlations with Soil Classification Systems

Both the AASHTO and Unified soil classification systems were devised for the specific purpose of assessing the suitability of soils for use in road and airfield construction. Since the CBR value of a soil is also a measure of its performance as a subgrade, logic suggests that there should be some general relationship between the soil groups and CBR values, and several such correlations do exist. Figures 7.6 and 7.7 give a compendium of correlations between CBR and Unified and AASHTO soil classes, based on recommendations by the US Highways Research Board and by the US Corps of Engineers (as presented by Liu 1967) and by Morin and Todor (1977) for South American red tropical soils.

7.3   CBR and Undrained Shear Strength

The CBR test can be thought of as a bearing capacity problem in miniature, in which the standard plunger acts as a small foundation. The most commonly used bearing capacity equations (Terzaghi 1943; Meyerhof 1951; Brinc Hansen 1970) all give the bearing capacity for a circular footing at the surface of a purely cohesive soil as:

$$q_u = 1.2cN_c \qquad (7.7)$$

where c is the cohesion and $N_c = 5.14$ (i.e. $2 + \pi$).

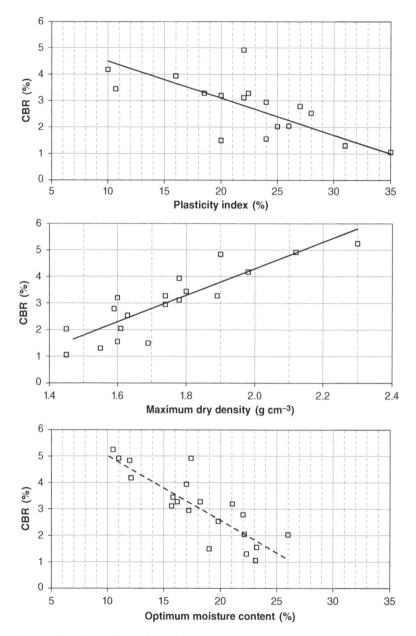

Figure 7.5 Correlations of plasticity index, maximum dry density and optimum moisture content with CBR. Adapted from Shirur and Hiremath (2014). Reproduced with permission of IOSR.

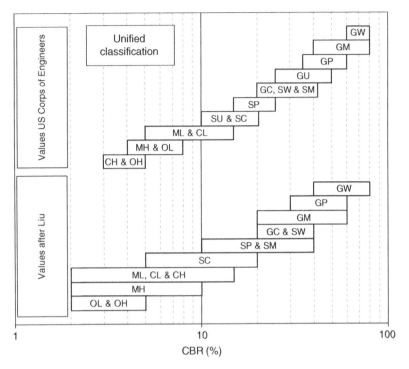

Figure 7.6   Approximate relationships between Unified soil classes and CBR values. Adapted from Liu (1967) and US Army Corps of Engineers (1970). Reproduced with permission of Transportation Research Board.

The equation therefore reduces to:

$$q_\mathrm{u} = 6.2c. \tag{7.8}$$

Using SI units, with q_u and c in kPa, the CBR value is 100% for a plunger pressure of 6900 kPa at a penetration of 2.5 mm, giving:

$$\mathrm{CBR} = \frac{q_\mathrm{u} \times 100}{6900} = 0.09c \tag{7.9}$$

or, as a rough rule-of-thumb, CBR = 0.1c.

Work carried out by Black (1961) on single-sized sand, and correlations with other work for clay, suggests that this approach gives calculated CBR

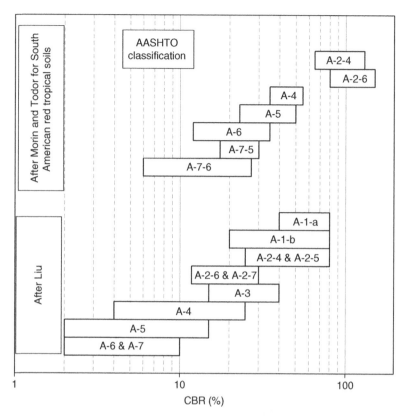

Figure 7.7 Approximate relationships between AASHTO soil classes and CBR values. Adapted from Liu (1967) and Morin and Todor (1977). Reproduced with permission of Transportation Research Board.

values that are close to measured values for field tests. While most correlations suggest a linear relationship between c and CBR, some suggest that the relationship is non-linear. For instance, Jenkins and Kerr (1998) suggest the relationship:

$$\text{CBR} = 0.0037c^{1.7}. \tag{7.10}$$

This is in broad agreement with our suggested correlation, equating to:

$$\text{CBR} = 0.076c \quad \text{for} \quad c = 75\text{kPa};$$
$$\text{CBR} = 0.12c \quad \text{for} \quad c = 150\text{kPa}; \quad \text{and}$$
$$\text{CBR} = 0.20c \quad \text{for} \quad c = 300\text{kPa}.$$

Laboratory CBR values can be expected to be a little higher because of the restraining influence of the mould. Black (1961) also suggests that, when calculating q_u, the substitution

$$c = s \tan\varphi_r \qquad (7.11)$$

be used, where φ_r is the true angle of internal friction.

7.4   An Alternative to CBR Testing

A significant drawback of CBR testing is that sample preparation and testing are time-consuming and even *in situ* testing is a fairly lengthy procedure, requiring a heavy vehicle as a kentledge. An alternative approach, especially popular in Europe, is to use a lightweight falling weight deflectometer to measure the modulus of subgrade reaction, E_v. The equipment is portable

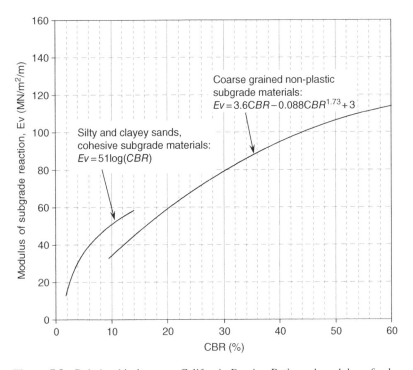

Figure 7.8   Relationship between California Bearing Ratio and modulus of subgrade reaction. Adapted from Defence Estates (2006).

and quick to use, allowing many more tests to be carried out. Various specifications exist (BS 2009; ASTM 2011) but of particular interest is the German standard (DIN 2012) which specifies taking two tests at each location, producing two E_v values, E_{v1} and E_{v2}, for the first and second tests, respectively, which are used to determine whether subgrade improvement is required. Further information from this is given by Fountain and Suckling (2012) and Tosovic et al. (2006). A relationship between modulus of subgrade reaction and CBR, given by Defence Estates (2006), is given in Figure 7.8. The equations for the curves were derived by the authors and added to the original plot, to aid use with spreadsheet calculations.

References

Agarwal, K. B. and Ghanekar, K. D. 1970. Prediction of CBR from plasticity characteristics of soils. *Proceedings of 2nd South-East Asian Conference on Soil Engineering,* Singapore, 571–576.

ASTM. 2011. *Standard test method for measuring deflections using a portable impulse plate loading test device.* American Society for Testing and Materials, International, E 2835-11.

Black, W. P. M. 1961. The calculation of laboratory and in-situ values of California bearing ratio from bearing capacity data. *Geotechnique* 11: 14–21.

Black, W. P. M. 1962. A method of estimating the CBR of cohesive soils from plasticity data. *Geotechnique* 12: 271–272.

Brinc Hansen, J. 1970. A revised and extended formula for bearing capacity. *Bulletin No. 28,* Danish Geotechnical Institute, Copenhagen, Denmark.

BS. 1990. *Methods of test for soils for civil engineering purposes.* British Standards Institution, BS 1377.

BS. 2009. *Code of practice for earthworks.* British Standards Institute, BS 6039.

de Graft-Johnson, J. W. S. and Bhatia, H. S. 1969. The engineering characteristics of the lateritic gravels of Ghana. *Proceedings of 7th International Conference on Soil Mechanics and Foundation Engineering,* Mexico, 2: 13–43.

de Graft-Johnson, J. W. S., Bhatia, H. S. and Yaboa, S. L. 1972. Influence of geology and physical properties on strength characteristics of lateritic gravels for road pavements. *Highway Research Board Record* 405: 87–104.

Defence Estates. 2006. *A Guide to Airfield Pavement Design and Evaluation, DMG27.* UK Ministry of Defence.

DIN. 2012. Determining the deformation and strength characteristics of soil by plate loading test. Deutsche Norm, DIN 18134: 2012-04.

Fountain, F. and Suckling, A. 2012. Reliability in the testing and assessing of piling work platforms. *Ground Engineering* November 2012: 29–30.

Jenkins, P. and Kerr, I.A. 1998. The strength of well-graded cohesive fill. *Ground Engineering* March 1998, 38–41.

Liu, T. K. 1967. A review of engineering soil classification systems. *Highway Research Board Record* 156: 1–22.

Meyerhof, G.G. 1951. The ultimate bearing capacity of foundations. *Geotechnique* 2(4): 301–332.

Ministry of Transport. 1969. Specification for road and bridge works. HMSO, London.

Morin, W. J. and Todor, P.C. 1977. Laterites and lateritic soils and other problem soils of the tropics. United States Agency for International Development A 1O/csd 3682.

Shirur, N. B. and Hiremath, S. G. 2014. Establishing relationships between CBR value and physical properties of soil. *IOSR Journal of Mechanical and Civil Engineering* 11(5): 26–30.

Terzaghi, K. 1943. *Theoretical Soil Mechanics*. John Wiley & Sons, New York.

Tosovic, S., Jovic, Z. and Nikolic, M. 2006. Classic and modern control of materials compaction during road construction with particular reference to the permanent control. *Slovenia Kongres o Cestah in Prometu*, Portoroz, October 2006, 1–7.

Transport and Road Research Laboratory. 1970. A guide to the structural design of new pavements. TRRL, Road Note 29, HMSO.

US Army Corps of Engineers. 1970. *Engineering and design: pavement design in frost conditions*. Corps of Engineers, EM-110-345-306.

8

Shrinkage and Swelling Characteristics

Expansive soils are those that show a marked volume change with increase or decrease in moisture content. Such swelling properties are restricted to soils containing clay minerals which are susceptible to penetration of their chemical structure by water molecules.

Clay swelling and consequential ground heave is a common annual phenomenon in areas where prevailing climatic conditions lead to significant seasonal wetting and drying, the greatest seasonal heave occurring in regions with semi-arid climates where pronounced short wet and long dry periods lead to major moisture changes in the soil. Moisture content changes may also result, in these regions and others, from human activity, such as the removal of vegetation, drainage, flooding and construction works.

8.1 Identification

The simplest swelling identification test is called the free-swell test (Holtz and Gibbs 1956). The test is performed by slowly pouring $10\,cm^3$ of dry soil fines ($<425\,\mu m$) into a $100\,cm^3$ graduated cylinder filled with water, and observing the equilibrium swelled volume. Free swell (%) is defined as:

Soil Properties and their Correlations, Second Edition. Michael Carter and Stephen P. Bentley.
© 2016 John Wiley & Sons, Ltd. Published 2016 by John Wiley & Sons, Ltd.

Table 8.1 Free swelling data for clay minerals.

Clay mineral and location of test specimens	Free swell (%)
Sodium montmorillonite	1400–1600
Calcium montmorillonite	65–150
Illite	60–120
Kaolinite	5–60*

*Published data shows free swell up to this value, but it is generally less than 20%.

Table 8.2 Typical Atterberg limit values for common clay minerals.

Clay mineral	Dominant pore water cation			
	Ca^{++}		Na^+	
	PL	LL	PL	LL
Montmorillonite	65–79	123–177	86–97	280–700
Illite	36–42	69–100	34–41	61–75
Kaolinite	26–36	34–73	26–28	29–52

$$\text{Free swell} = \frac{(\text{final volume}) - (\text{initial volume})}{(\text{initial volume})} \times 100. \qquad (8.1)$$

Table 8.1 gives free swelling values, measured on US soils, for some common clay minerals.

In field situations the amount of swelling or shrinkage, or whether any volume change occurs at all, will depend on a number of factors such as moisture content changes, thickness of the deposit, initial density, groundwater chemistry and confining pressures. However, a fundamental ingredient is commonly the presence of montmorillonite or other swelling mineral and, more specifically, its proportion in the soil. In some instances clay-mineral type can be identified from the origin and geological setting of the soil, together with consideration of Atterberg limits. Typical ranges of Atterberg limits are shown in Table 8.2; note the effect of the dominant cation in the pore water. Another indicator of clay mineral type is Skempton's (1953) activity (A_c) which relates plasticity index to the proportion of clay present in the soil, defined as:

$$A_c = \frac{PI}{C} \qquad (8.2)$$

where C is the percentage finer than 0.002 mm. Typical activity values are:

sodium montmorillonite: 7.2;
calcium montmorillonite: 1.5;
illite: 0.9; and
kaolinite: 0.33–0.46.

8.2 Swelling Potential

It is generally recognised that the best indication of the susceptibility of a soil to shrinkage or swelling due to decreases or increases in moisture content is provided by the swelling potential test.

Swelling potential is defined as the percentage swell of a laterally confined sample that has been compacted to maximum density at optimum moisture content according to the standard compaction test (BS 1377:1990, Test 12, 2.5 kg rammer method; or AASHTO 1982, T99, 5.5 lb rammer method) and then allowed to swell under a surcharge of 6.9 kPa (1 lb/in²). The test is carried out in a standard oedometer (consolidometer) as for consolidation testing.

In order to give meaning to the significance of swelling potential values, descriptive terms are used for various ranges of swelling potential as indicated in Table 8.3.

8.2.1 Swelling Potential in Relation to other Properties

The swelling potential test, including specimen preparation and testing, is quite lengthy and is not normally carried out; swelling potential being inferred from other properties. A number of researchers have tried to correlate swelling potential with plasticity index. Since both the liquid and plastic limits and the swelling properties of a soil are governed by the amounts and types of clay minerals present, it seems reasonable that such a correlation should exist. Seed *et al.* (1962) established the relationship:

$$S = 60K \times \mathrm{PI}^{2.44} \tag{8.3}$$

Table 8.3 Descriptive terms for swelling potential.

Swelling potential (%)	Description
0–1.5	Low
1.5–5.0	Medium
5.0–25	High
25+	Very high

Table 8.4   Identification of swelling soils based on plasticity index.

Swelling potential (%)	Plasticity index[1]		Modified plasticity index[2,3]	
	Seed *et al.* (1962)	Krebs and Walker (1971)	Building Research Establishment (1993)[4]	National House-Building Council (2008)
Low	0–15	0–15	<20	<20
Medium	10–30	15–24	20–40	20–40
High	20–55	25–46	40–60	>40
Very high	>40	>46	>60	–

[1] Swelling potential descriptions for columns 2 and 3 are as defined in Table 8.3.
[2] Modified plasticity index used by the BRE and NHBC are equivalent to plasticity index × proportion passing through a 425 μm sieve.
[3] The swelling categories given in columns 4 and 5 by the BRE and NHBC are not defined in terms of numerical swelling potential.
[4] BRE use the term 'volume change potential' to avoid confusion with the NHBC guidelines.

where S is the swelling potential; PI is the plasticity index and K is a constant, equal to 3.6×10^{-5}.

This equation applies to soils with clay contents of between 8% and 65%. The calculated value is probably accurate to within about 33% of the laboratory value. Although their results are based on work with artificial mixtures of sands and clays, the correlation has been shown to be applicable to natural soils. Using this equation and allowing for the possible 33% error in calculated values of swelling potential, ranges of plasticity index values may be obtained for the various classes of swelling potential, as indicated in columns 1 and 2 of Table 8.4. Also indicated in the table (column 3) are values suggested by Krebs and Walker (1971).

It seems reasonable to assume that the swelling potential of a soil depends not only on the swelling properties of the fines but also on the proportion of fines: intuitively, a soil that consists entirely of fines will swell more than a soil with similar fines but with, say, 50% coarse material. Recognising this, the UK Building Research Establishment (BRE 1993) and the UK National House Building Council (NHBC 2008) changed their swelling potential categories from being based solely on plasticity index to being based on modified plasticity index, as defined in the notes below Table 8.4. Both the BRE and NHBC swell categories are used in conjunction with design methods to determine minimum depths for foundations near trees, depending on the soil swelling potential, climatic factors and proximity and species of trees. The BRE also gives guidance on the design of piled foundations in potentially expansive clays.

A correlation between swelling potential and plasticity index was found by Chen (1988), based on tests of 321 undisturbed samples. He proposed:

$$S = Be^{A \times PI} \tag{8.4}$$

where $B = 0.2558$; $A = 0.0838$; and e is the base of the natural logarithms, 2.718. Chen also established a correlation of plasticity index against a swelling potential obtained for a surcharge pressure of 48 kPa. A comparison of various correlations between swelling potential and plasticity index is shown in Figure 8.1, from which it can be seen that there are wide variations between the findings of different researchers. It should be noted that the Holtz and Gibbs (1956) correlation given in the figure is not really comparable with the others since their volume change measurements were carried out on air-dried specimens of undisturbed soil. Their values given in the chart are therefore not strictly swelling potential. This is discussed later in this section.

Although soils exhibiting high swelling characteristics usually have high plasticity indices, not all soils with high plasticity indices have a high swelling potential. The plasticity index can therefore be used only as a rough guide to swelling potential. Logic suggests that there should be relationships between potential for expansion and both shrinkage limit and linear shrinkage. Table 8.5 shows a general guide for these relationships suggested by Altmeyer (1955). However, although a knowledge of shrinkage limit is useful in assessing potential volume changes, other researchers have been unable to establish a conclusive correlation between it and swelling potential (Chen 1988). The fact that measured shrinkage limit values vary significantly according to the test method, as noted in Chapter 2, gives further doubt about the value of shrinkage limit as a predictor of swelling potential.

Work by Seed et al. (1962) suggests that there is a correlation between swelling potential and the content of clay-sized particles (finer than 0.002 mm). Unfortunately, the correlation includes factors which depend on the type of clay present, which limits its usefulness for normal site investigations. They therefore suggested an alternative approach using the concept of activity. Swelling potential is related to activity as shown in Figure 8.2. However, they suggest that when using this figure, Skempton's definition of activity be modified to:

$$A_c = \frac{PI}{C - 5} \tag{8.5}$$

where PI is the plasticity index and C is the percentage finer than 0.002 mm.

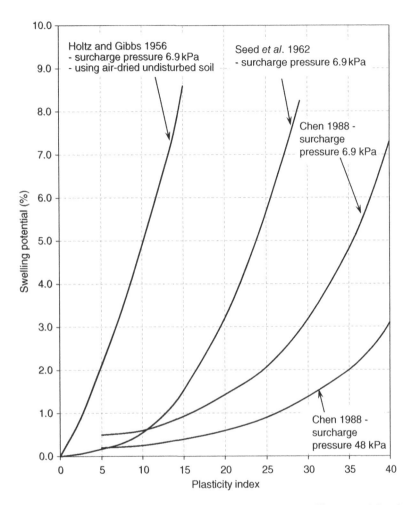

Figure 8.1 A comparison of various correlations between swelling potential and plasticity index. Adapted from Chen (1988). Reproduced with permission of Elsevier.

Table 8.5 Suggested guide to the determination of potential for expansion using shrinkage limit and linear shrinkage. Adapted from Altmeyer (1955).

Potential for expansion	Shrinkage limit (%)	Linear shrinkage (%)
Critical	<10	>8
Marginal	10–12	5–8
Non-critical	>12	<5

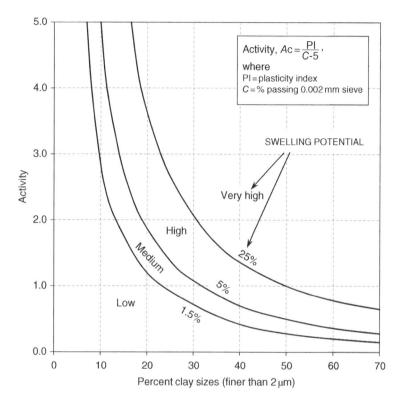

Figure 8.2 Classification chart for swelling potential. Adapted from Seed *et al.* (1962). Reproduced with permission of American Society of Civil Engineers.

This is because a plot of plasticity index against clay content passes through the origin for clay contents in excess of 40% but not for lower clay contents, as indicated in Figure 8.3. Using the amended definition helps to compensate for this for soils with the lower clay contents.

Holtz and Gibbs (1956) correlated volume change with colloid content (defined as finer than 0.001 mm), plasticity index and shrinkage limit, as indicated in Figure 8.4. They suggest that, because of the uncertainty of the correlations, the potential for expansion should be assessed by the simultaneous consideration of all three correlations, as indicated in Table 8.6. Their procedure has been adopted by the US Water and Power Resources Service (formerly the US Bureau of Reclamation). It should be remembered that their volume change measurements, whilst being made at a pressure of 6.9 kPa are for air-dried undisturbed soils and so are not directly

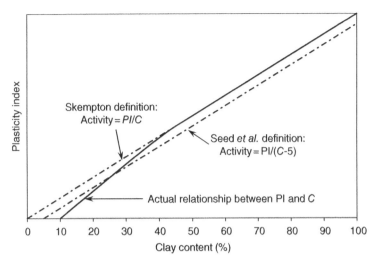

Figure 8.3 Simplified relationship between plasticity index and clay content. Adapted from Seed *et al.* (1962). Reproduced with permission of American Society of Civil Engineers.

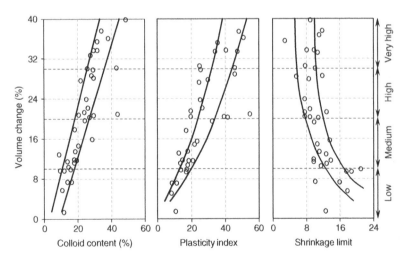

Figure 8.4 Relationships between volume change and colloid content, plasticity index and shrinkage limit, for air-dry to saturated conditions under a load of 6.9 kPa. Adapted from Holtz and Gibbs (1956). Reproduced with permission of American Society of Civil Engineers.

Table 8.6 Estimation of potential volume changes of clays. Adapted from Holtz and Gibbs (1956). Reproduced with permission of American Society of Civil Engineers.

Colloid content (% finer than 1 μm)	Data from index tests		Probable expansion* (% total volume change)	Potential for expansion
	Plasticity index	Shrinkage limit		
>28	>35	<11	>30	Very high
20–31	25–41	7–12	20–30	High
13–23	15–28	10–16	10–30	Medium
<15	<18	<10	<10	Low

*Based on a loading of 6.9 kPa.

comparable with the values of swelling potential discussed previously (see Figure 8.1). Also, their results are based on only 45 samples.

A more sophisticated relationship which can take into account the change in moisture content from an initial value to saturation is presented by Weston (1980). This correlation, established for soil in the Transvaal, is essentially a more fully developed version of previous relationships described by Williams (1957) and Van de Merwe (1964). The amount by which a soil could potentially swell is given by:

$$\text{Swell}(\%) = 0.00041 \times W_{LW}^{4.17} P^{-0.386} w_i^{-2.33} \qquad (8.6)$$

where w_i is the initial water content, P is the vertical pressure under which swell takes place and W_{LW} is the weighted liquid limit, defined by:

$$W_{LW} = \text{LL} \times \frac{\% \text{ of material finer than } 425\,\mu m}{100} \qquad (8.7)$$

where LL is the liquid limit.

Regardless of their propensity to swell, soils that are subjected to inundation for the first time may either swell or collapse, depending on their state of compaction; that is, their dry density. Such a situation may occur where, for instance, a quarry fill is flooded for the first time.

Data given by Holtz and Kovacs (1981), Holtz et al. (2011) and others suggests that the critical dry density dividing collapse and swell depends on the liquid limit. The critical density (kN/m³) that divides collapse from swelling may be taken as:

$$Y_{d(crit)} = 14.2 - 35.5 \log \text{LL.} \qquad (8.8)$$

Close to this density, neither excessive swelling nor collapse is anticipated. Collapse is increasingly likely at densities significantly below this value, while swelling is possible for expansive soils with densities significantly above this value. It should be remembered, however, that this greatly simplifies a complex interaction between density and swelling potential, so this method should be used only for initial assessments, subject to checking by consolidation testing.

8.2.2 *Reliability of Swell Predictions Based on Correlations*

As noted above, the actual swell that will occur in a soil depends on the environmental conditions under which it exists, principally changes in moisture content, so that laboratory testing can do no more than give an indication of its propensity to swell or shrink, but it is generally acknowledged that the most reliable indicator of swelling is swell potential as measured in the oedometer test. It will also be seen from the various comments in the previous section that there is significant doubt about the reliability of swelling potential predictions based on other properties.

A review of the various correlations by Sridharan and Prakash (2000) using both kaolinitic and montmorillonitic soils showed that correlations based on liquid limit values, plasticity index values and activity all tended to overestimate the swelling potential. However, they did find reasonable agreement between swelling potential and free swell index, although they used a modified form of free swell index test to overcome problems of the standard procedure which can give negative values with some kaolin-rich soils. They note the work of previous researchers, notably Chen, which indicates that while high-swelling soils tend to have high liquid limits and plasticity indices, not all soils with these high index properties are high swelling.

In summary, it seems that the propensity of a clay to swell can be reliably determined only by direct measurement of its swelling properties, notably in the swell potential test and, to a slightly lesser extent, the free swell test. Notwithstanding their wide acceptance in standards, correlations based on other properties give no more than a very rough indication of swelling potential.

8.3 Swelling Pressure

Once a potentially expansive soil has been identified and a qualitative indication of the potential swell has been made, an evaluation of the swelling pressure is necessary for design purposes, this being the pressure that would be exerted against a rigid buried structure. Swelling pressure can be determined from a one-dimensional oedometer (consolidometer) test: a number of variations of this test have been developed (Jennings and Knight 1957;

Burland 1975), but commonly the specimen is flooded and the load required to maintain constant volume is recorded (Fredlund 1969). Alternatively, the swelling pressure can be estimated from empirical relationships with more routinely-measured parameters. Komornik and David (1969) suggested a means of estimating the swelling pressure using a swell index I_s, defined as:

$$I_s = \frac{m}{LL} \qquad (8.9)$$

where m is the natural moisture content and LL is the liquid limit.

The relationship between I_s and swelling pressure across a range of liquid limits is shown in Figure 8.5. Based on experience with expansive soils in

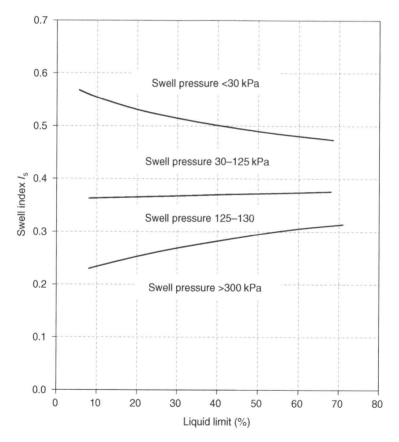

Figure 8.5 Relationship between swell index and swelling pressure for a range of liquid limits. Adapted from Komornik and David (1969). Reproduced with permission of American Society of Civil Engineers.

Table 8.7   Estimating probable swelling pressure. Adapted from Chen (1988).
Reproduced with permission of Elsevier.

Laboratory and field data			Probable expansion (% total volume)	Swelling pressure (kPa)	Degree of expansion
Percentage passing 75 μm sieve	Liquid limit	SPT N-value			
>35	>60	>30	>10	>1000	Very high
60–95	40–60	20–30	3–10	250–1000	High
30– 60	30–40	10–20	1–5	150–250	Medium
<30	<30	<10	<1	<50	Low

the Rocky Mountain area of the United States, Chen (1988) suggested a predictive relationship for swelling pressure using percentage of fines, liquid limit and the standard penetration resistance, as given in Table 8.7. Note that the 'probable expansion' given in this table is the swelling potential for a confining load of 48 kPa ($1000\,lb/ft^2$), based on the premise that this is a typical foundation pressure for light structures.

A number of theoretical equations have been developed for computing heave in expansive soils. Most require an evaluation of the swelling pressure (Fredlund *et al.* 1980; Rama Rao and Fredlund 1980) but some are based on measurement of soil suction (Johnson 1980; Snethen 1980).

References

AASHTO. 1982. *Standard specifications for transportation materials and methods of testing and sampling.* American Association of State Highway and Transportation Officials. Washington, DC, USA.

Altmeyer, W. T. 1955. Discussion of engineering properties of expansive clays. *Proceedings of ASCE, Journal of Soil Mechanics and Foundation Division* 81: 17–19.

BRE. 1993. *Low-rise buildings on shrinkable clay soils.* Building Research Establishment, Digest 240.

BS. 1990. *Methods of test for soils for civil engineering purposes.* British Standards Institution, BS 1377.

Burland, J. B. 1975. Discussion. *Proceedings of 6th Regional Conference for Africa on Soil Mechanics and Foundation Engineering* I: 168–169.

Chen, F. H. 1988. *Foundations on Expansive Soils.* Elsevier, Amsterdam, Developments in Geotechnical Engineering, no. 12, 280 pp.

Fredlund, D. G. 1969. Consolidometer test procedural factors affecting swell proper ties. *Proceedings of 2nd International Conference on Expansive Clay Soils.* Texas, 435–456.

Fredlund, D. G., Hasan, J. U. and Filson, H. L. 1980. The prediction of total heave. *Proceedings of 4th International Conference on Expansive Soils,* Denver, Colorado, 1–17.

Holtz, R. D. and Kovacs, W. D. 1981. *An Introduction to Geotechnical Engineering.* Prentice-Hall, New Jersey, 733 pp.

Holtz, R. D., Kovacs, W. D. and Sheahan, T. C. 2011. *An Introduction to Geotechnical Engineering.* Pearson, New Jersey, USA.

Holtz, W. G. and Gibbs, H. J. 1956. Engineering properties of expansive clays. *Transactions ASCE* 121: 641–677.

Jennings, J. E. B. and Knight, K. 1957. The prediction of total heave from the double oedometer test. *Transactions of South African Institute of Civil Engineers* 7: 285–291.

Johnson, L. D. 1980. Field test sections on expansive soil. *Proceedings of 4th International Conference on Expansive Soils,* Denver, Colorado.

Komornik, A. and David, D. 1969. Prediction of swelling pressure of clays. *Journal of Soil Mechanics, ASCE* 95(1): 209–226.

Krebs, R. D. and Walker, E. D. 1971. *Highway Materials.* McGraw-Hill, New York.

NHBC. 2008. *Building Near Trees.* National House-Building Council, Milton Keynes. Chapter 4.2.

Rama Rao, R. and Fredlund, D.G. 1988. Closed-form heave solutions for expansive soils. *Proceedings ASCE Journal of Geotechnical Engineering Division* 114: 573–588.

Seed, H. B., Woodward, R. J. and Lundgran, R. 1962. Prediction of swelling potential of compacted clays. *Proceedings ASCE Journal of Soil Mechanics and Foundation Division* 88: 107–131.

Skempton, A. W. 1953. The colloidal activity of clays. *Proceedings of 3rd International Conference on Soil Mechanics and Foundation Engineering.* Zurich, 57–61.

Snethen, D. R. 1980. Characterization of expansive soils using soil suction data. *Proceedings of 4th International Conference on Expansive Soils,* Denver, Colorado, I: 54–75.

Sridharan, A. and Prakash, K. 2000. Classification procedures for expansive soils. *Proceedings of the Institution of Civil Engineers, Geotechnical Engineering* 143: 235–240.

Van der Merwe, D. H. 1964. Prediction of heave from the plasticity index and percentage clay fraction of soils. *Civil Engineer in South Africa* 6: 103–107.

Weston, D. J. 1980. Expansive roadbed treatment for Southern Africa. *Proceedings of 4th International Conference on Expansive Soils,* Denver, Colorado, I: 339–360.

Williams, A. A. B. 1957. Discussion. *Transactions of South African Institute of Civil Engineers* 8: 36.

US Bureau of Reclamation. 1974. *Earth Manual.* US Bureau of Reclamation, Denver, 810 pp.

9

Frost Susceptibility

Two potentially damaging effects are associated with frost action in soils: the expansion and lifting of the ground in winter (frost heave) and the loss of bearing capacity during the spring thaw. Soils that display one or both of these manifestations are referred to as 'frost susceptible'. The problem of frost damage is widespread; it occurs in temperate regions where there is seasonal soil freezing as well as in the high-latitude regions.

9.1 Ice Segregation

Although water increases in volume by about 9% on freezing, simple freezing of interstitial water causes little ground uplift. Frost heave occurs to a much greater extent where water is free to enter the soil and migrate to the freezing front where layers of ice grow parallel to the ground surface by displacing the overlying soil layer. The migrating water must come largely from groundwater below the layer in which ice is forming and segregating from the soil, since ice and frozen ground will effectively prevent any downward percolation from the ground surface.

The thermodynamics of moisture movement to the freezing front are complex; useful summaries are given by Chamberlain (1981), Harris (1987) and Davis (2001). The soil must be sufficiently permeable to allow water to

Soil Properties and their Correlations, Second Edition. Michael Carter and Stephen P. Bentley.
© 2016 John Wiley & Sons, Ltd. Published 2016 by John Wiley & Sons, Ltd.

flow through the soil but this alone is not sufficient to cause the extent of heave that occurs in frost-susceptible soils. For this, conditions must exist which allow water to migrate to the freezing front, enabling crystals and lenses of ice to grow (Rempel 2007). This can occur only where soil suction and surface tension effects within the soil capillaries lower the freezing point around particles, creating pathways for the migrating water to reach the freezing front (Figure 9.1). Soil suction and surface tension are related to the size of the pores between particles, hence to the particle sizes. Clay mineralogy is also a factor, since water within the adsorption complex of clay particles (Chapter 2) tends to have a lower freezing point.

Considering all the above requirements for ice lens formation, there is clearly a trade-off between soil suction, allowing migration of liquid water through the forming ice front, which requires small pores, and permeability, which requires larger pores. This is shown schematically in Figure 9.2.

There are many factors that affect frost susceptibility, including clay mineralogy and the number of freeze-thaw cycles, but four factors are of particular significance in affecting the amount of ice segregation during soil freezing: the pore size of the soil; the moisture supply; the rate of heat extraction; and the confining pressure. Theory and observation indicate that the suction potential of soils and their susceptibility to ice segregation increases as pore size decreases. However, as discussed above, the low permeability of heavy clays may restrict water migration sufficiently to prevent significant ice segregation (Penner 1968). Thus, highly frost-susceptible

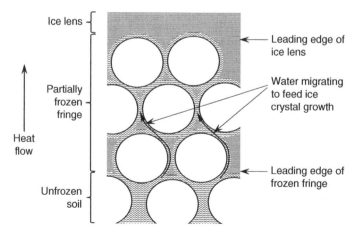

Figure 9.1 Mechanisms of ice lens formation.

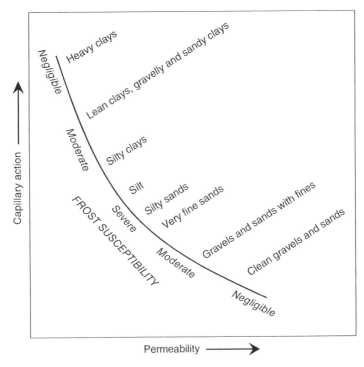

Figure 9.2 The influence of permeability and capillary action on frost susceptibility of soils.

soils possess pore size distributions which produce an optimum combination
of soil suction and permeability. In view of the close correspondence between pore size and grain size, and the relative ease with which the latter
may be measured, frost susceptibility criteria based on grain size are
frequently used.

9.2 Direct Measurement of Frost Susceptibility

The test that is usually considered to be the defining measurement of frost
susceptibility is the frost heave test which essentially consists of freezing a
soil specimen in a mould by cooling it at the top face, therefore applying a
temperature gradient, and measuring the amount of frost heave. Soil strength
after thawing may also be measured, often in the form of a CBR test (see
Chapters 1 and 7). However, many variations of test methods are used by

different agencies including specimen dimensions, the temperature gradient applied, the number of freeze-thaw cycles, surcharge pressure on the specimen, methods to reduce friction on the mould sides and the amount of insulation of the specimen sides. A commonly used version of the test in the US and other countries is that defined in ASTM D 5918 (2013); see also Roe and Webster (1984) and BS 812-124 (2009).

However, frost heave tests have a number of drawbacks; principally that they are costly to carry out and are time-consuming, tests lasting for typically ten days to several weeks, which makes them undesirable for use in large linear projects such as roads and pipelines where a large number of evaluations of frost susceptibility will usually be needed along the alignment. Further, reliance on frost heave alone may not be sufficient since soils may exhibit little heave but still show appreciable weakening on thawing, as described above.

9.3 Indirect Assessment of Frost Susceptibility

To overcome these problems, methods have been developed that rely on more standard testing to identify frost-susceptible soils, principally grading, moisture content and plasticity testing. A very large number of such methods exist; indeed, Chamberlain (1981) identified over 100 such methods. Unfortunately, this large number is an indication that most methods are either unsatisfactory or are suitable only for the specific soil and climatic conditions of the region for which they were developed. Some of the methods used are outlined in the following sections.

9.3.1 Grading

There are essentially two considerations that are used to identify frost susceptible materials from particle size characteristics: the overall grading and the percentage of fines; fines usually being defined as material passing the 0.075 mm sieve, although the <0.02 mm fraction is considered important by many agencies and researchers.

Essentially, coarse and medium sands are generally non-frost susceptible; that is, ice lenses do not normally develop when they freeze, whereas fine sands, silts and silty and sandy clays are frost susceptible and are subject to considerable ice lensing during freezing, providing a water supply is present. Heavy clays are typically not frost susceptible because their very low permeability inhibits the flow of water to the freezing front. This is summarised schematically in Figure 9.2.

The percentage of fines that need to be present for a soil to be considered frost susceptible is typically about 3%, but this limit varies widely according to agency or researcher. For instance, Glossop and Skempton (1945) observed that poorly graded (well sorted) soils in which less than 30% of the particles are silt size do not exhibit frost heaving characteristics. Casagrande (1932) suggested that the particle size critical to soil heave is 0.02 mm: if the proportion of such particles is less than 1%, no heave is expected, but considerable heaving may occur if this amount is over 3% in well-graded soils and over 10% in poorly graded soils. The influence of the <0.02 mm fraction was also demonstrated by Kaplar (1970) for gravelly sands where the coarser fraction was progressively removed. Figure 9.3 shows the relationship between average rate of heave (mm/day) and the percentage finer than 0.02 mm. However, these results were obtained under specific laboratory conditions and they should only be used as a general guide to the field response.

Examples of many grading limits used to identify frost-susceptible soils have been summarised by Chamberlain (1981), three examples of which are

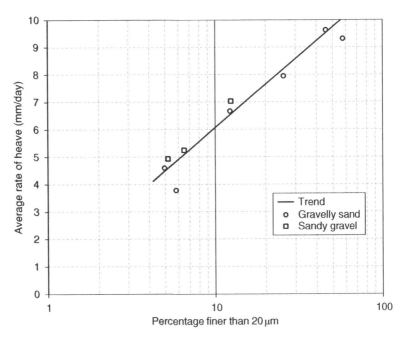

Figure 9.3 Relationship between the average rate of frost heave and the percentage of soil particles finer than 20 μm. Adapted from Kaplar (1970). Reproduced with permission of Transportation Research Board.

shown in Figure 9.4. The countries in which these were used are also indicated, although grading alone is no longer usually used; for instance, the UK Highways England Specification for Roads and Bridges (Highway Agency 2001) specifies frost susceptibility in terms of the freeze-thaw test. Nevertheless, the curves may be used for preliminary identification of potentially problematic soils. Many other such curves exist, with various limits. In some systems, the grading or average size of the fines is also considered to be an indicator of the degree of frost susceptibility.

Reed *et al.* (1979) noted that predictions from grain size distributions failed to take account of the fact that soils can exist at different states of density and therefore porosity, yet they have the same grain size distribution. They derived expressions for predicting frost heave Y in mm/day, and one of their simpler expressions, based on pore diameters, is:

$$Y = 1.694\left(\frac{O_{40}}{O_{80}}\right) - 0.3805 \qquad (9.1)$$

where O_{40} and O_{80} are the pore diameters whereby 40% and 80% of the pores are larger, respectively.

9.3.2 Plasticity

Frost susceptibility tends to be a feature of silty and sandy clays, that is, soils of low to medium plasticity. Table 9.1 gives a correlation of frost susceptibility and permeability with grading and plasticity index suitable for preliminary identification based on recommendations by the UK Transport and Road Research Laboratory (1970). A similar early classification system used by the US Army Corps of Engineers to assess frost susceptibility for pavement design (Table 9.2), involving grading and plasticity, was reported by Linell *et al.* (1963). Once again the critical particle size is given as 0.02 mm. The groups are in order of increasing frost susceptibility, with group F4 soils being particularly frost susceptible. A graphical relationship presented by the US Army Corps of Engineers (1965) showing the average rate of heave (mm/day) for a range of soil groups, defined by the Unified system, is given in Figure 9.5. A revised version of this was given by Kaplar (1974), but the earlier Corps of Engineers version is presented here as this is more commonly used. It can be seen from the chart that there is a large variability in frost susceptibility, in terms of rate of heave, within groups.

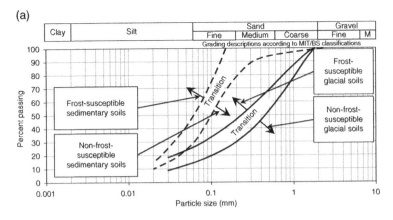

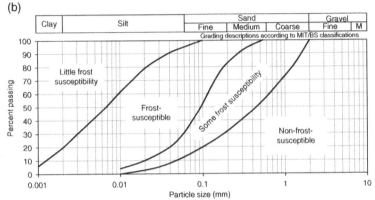

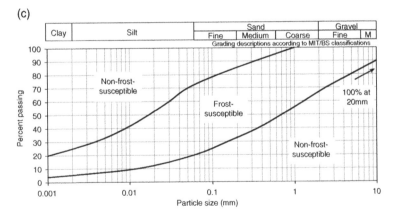

Figure 9.4 A selection of proposed limits of frost susceptibility based on grading according to (a) Beskow (1935) for Denmark, Greenland and Sweden; (b) Armstrong and Csathy (1963) for Canada; and (c) Croney (1949) for UK. Adapted from Chamberlain (1981).

Table 9.1 Preliminary identification of frost susceptible soils. Adapted from TRRL (1970).

Permeability rating	Identification	Frost susceptibility
High permeability	Granular: <10% finer than 75 μm	Not susceptible
Intermediate permeability	Granular: >10% finer than 75 μm; cohesive: PI <20	Susceptible
Low permeability	Cohesive: PI > 20	Not susceptible

Table 9.2 Frost susceptibility of soils related to soil classification. Adapted from US Army Corps of Engineers (1965).

Group	Region of applicability
NFS	All soils with less than 3% finer than 20 μm
F1	Gravelly soils: 3–20% finer than 20 μm
F2	Sands: 3–15% finer than 20 μm
F3	(a) Gravelly soils: >20% finer than 20 μm
	(b) Sands (except silty fine sands): >15% finer than 20 μm
	(c) Clays: PI > 12
	(d) Varved clays: with uniform conditions
F4	(a) Silts: including sandy silts
	(b) Fine silty sands: >15% finer than 20 μm
	(c) Lean clays: PI < 12
	(d) Varved clays: with non-uniform conditions

Frost susceptibility increases from group F1 to group F4. NFS: non-frost-susceptible.

Migration of water and frost heaving are also influenced by the mineralogy of the clay fraction. Clay minerals with expandable structures are able to hold more water, but the water is relatively immobile compared with non-expandable clay minerals. Consequently, strong frost heaving is more likely to be associated with soils where the fines are devoid of montmorillonite and related minerals so that, for example, clayey soils containing kaolin tend to exhibit higher frost susceptibility than clayey soils containing montmorillonite.

9.3.3 Predictions Based on Segregation Potential

The approaches discussed above, whist commonly used, all predict frost susceptibility in a general way, simply identifying frost-susceptible soils.

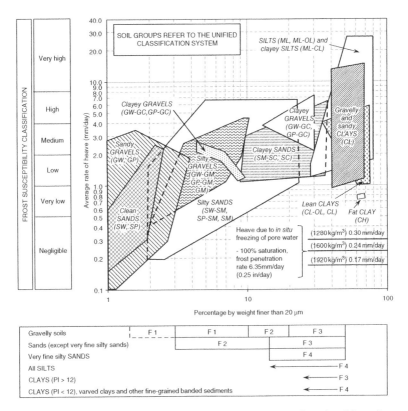

Figure 9.5 Average rate of heave plotted against percentage finer than 20 μm from laboratory tests of a range of natural soils. Adapted from US Army Corps of Engineers (1965).

The concept of segregation potential, in comparison, promises a method of making specific predictions of heave.

As discussed previously, frost susceptibility, both heave on freezing and weakening on thawing, is caused by the formation of segregated ice crystals. The propensity for ice crystals to form in a given soil should therefore be an indication of its frost susceptiblility. Konrad and Morgenstern (1981) suggested a new approach using this logic to estimate frost susceptibility based on the concept of segregation potential, SP. This is obtained from specialised step-freezing tests and from field observations, making the determination of SP values slow and expensive, so unsuitable for large projects such as highways and pipelines. To overcome this problem, Konrad (1999) proposed a method of predicting SP from a combination of: the 50% (D_{50}) particle size of the fines (<75 μm); the liquid limit; the water content; and the specific surface of the fines. These are standard tests except for the specific surface test, which is more specialised. The value of SP allows, in theory, specific predictions to be made about heave. However, this requires both a knowledge of the freezing index for the site, as the number of degree-days, and either specialist knowledge of heave calculations or dedicated software to give a numerical solution. Thus, while the method promises to be a predictive tool for the future, it is likely to be suitable only for engineers with specialist knowledge of frost heave calculations and would need validating for a much wider range of soils than those investigated by Konrad.

9.4 Choosing a Suitable Method of Evaluating Frost Susceptibility

It will be seen from the above discussions that most established methods of evaluating frost susceptibility do so in a general way, simply classifying the degree to which materials are frost susceptible without attempting to quantify actual amounts of frost heave or weakening on thawing. Moreover, methods are often based on early work in this field, sometimes dating back to as early as the 1930s. While some more recent studies do incorporate predictions of heave or weakening, such as the segregation potential method described above, these methods are not easy to apply in practice. A further characteristic of the methods is that they tend to be based on information from specific regions and climatic conditions and, when used for other regions, geological conditions or climates, their predictions may be misleading. Indeed, that the classification systems vary so widely, for instance in specified grading or plasticity limits, implies that frost susceptibility criteria are not readily transferable outside the region, geology and climate for which they were developed.

Given the above factors, the most appropriate way to assess frost susceptibility would seem to be to use a method developed for the specific region, climate and soils. If methods developed elsewhere are used (because, for instance, they are thought to be more comprehensive or specific in their predictions), they should be compared with predictions of an established local method and preferably also with actual measurements or experience.

References

Armstrong, M. D. and Csathy, T. I. 1963. Frost design practice in Canada. *Highway Research Record* 33: 170–201.

ASTM. 2013. *Standard method of test for frost heave and thaw weakening susceptibility of soils*. American Society for Testing and Materials, D 5918. West Conshohocken, Pennsylvania, USA.

Beskow, G. 1935. *Soil Freezing and Frost Heaving with Special Application to Roads and Railways*. Swedish Geotechnical Society, Stockholm, Series C, no. 375.

BS. 2009. *Testing aggregates: method for determination of frost heave*. British Standards Institute, 812-124.

Casagrande, A. 1932. Research on the Atterberg limits of soils. *Public Roads* 13(8): 121–136.

Chamberlain, E. J. 1981. *Frost Susceptibility of Soil: Review of Index Tests*. US Army Corps of Engineers, Hanover. CRREL Monograph 81–2, 110 pp.

Croney, D. 1949. *Some Cases of Frost Damage to Roads*. Transport and Road Research Laboratory, HMSO, London, Road Note 8.

Davis, N. 2001. *Permafrost: A Guide to Frozen Ground in Transition*. University of Alaska Press, Fairbanks, Alaska.

Glossop, R. and Skempton, A. W. 1945. Particle size in silts and sands. *Journal of the Institute of Civil Engineers* 25: 81–105.

Harris, C. 1987. Mechanisms of mass movement in periglacial environments. In: *Slope Stability* (eds M. G. Anderson and K. S. Richards). John Wiley, 531–560.

Highway Agency. 2001. *Pavement Design and Construction*. Design Manual for Roads and Bridges, 7(2). Highway Agency, UK.

Kaplar, C. W. 1970. Phenomenon and mechanism of frost heaving. *Highway Research Record* 304: 1–13.

Kaplar, C. W. 1974. *Freezing Test for Evaluating Relative Frost Susceptibility of Various Soils*. US Army Corps of Engineers, Hanover. CRREL Technical Report, 223.

Konrad, J-M. 1999. Frost susceptibility related to soil index properties. *Canadian Geotechnical Journal* 36: 403–417.

Konrad, J-M. and Morgenstern, N. R. 1981. The segregation potential of a freezing soil. *Canadian Geotechnical Journal* 18: 482–491.

Linell, K. A., Hennion, F. B. and Lobacz, E. F. 1963. Corps of Engineer's pavement design in areas of seasonal frost. *Highway Research Record* 33: 76–136.

Penner, E. 1968. Particle size as a basis for predicting frost action in soils. *Soils and Foundations* 8: 21–29.

Reed, M. A., Lovel, C. W., Altschaeffi, A.G. and Wood, L. E. 1979. Frost heaving rate predicted from pore size distribution. *Canadian Geotechnical Journal* 16: 463–472.

Rempel, A. W. 2007. Formation of ice lenses and frost heave. *Journal of Geophysical Research* 112: 1–17.

Roe, P. G. and Webster, D. C. 1984. *Specification for the TRRL Frost Heave Test*. Transport and Road Research Laboratory, HMSO, London, Special Report no. 829, 39 pp.

Transport and Road Research Laboratory, 1970. *A Guide to the Structural Design of New Pavements*. Transport and Road Research Laboratory, HMSO, London, Road Note no. 29.

US Army Corps of Engineers 1965. Soils and geology – pavement design for frost conditions. Department of the Army Technical Manual, TM 5-818-2. US Army Corps of Engineers, Hanover.

10

Susceptibility to Combustion

Some materials, especially old colliery discard material (coal tips), may contain sufficient carbonaceous material to present a risk of combustion. Combustion in old tips can arise in two ways:

- material near the surface may catch fire as a result of surface fires started either accidentally or deliberately, such as the burning of refuse; and
- material deep within the tip can spontaneously combust as a result of heat generated by the gradual degradation of carbonaceous material within the tip.

In both cases the risk depends on the proportion of combustible material present and, in the case of deep spontaneous combustion, the proportion of air voids that are a source of oxygen, allowing burning to take place. Remedial measures include covering near-surface material to insulate it from the heat of any surface fires and excavating deeper, loosely placed material and re-depositing it in well-compacted layers to reduce the air voids content.

The potential for combustion is normally assessed by measuring the calorific value of the material, and guidance is given in terms of the calorific value test. The ICRCL (1986) states that:

Soil Properties and their Correlations, Second Edition. Michael Carter and Stephen P. Bentley.
© 2016 John Wiley & Sons, Ltd. Published 2016 by John Wiley & Sons, Ltd.

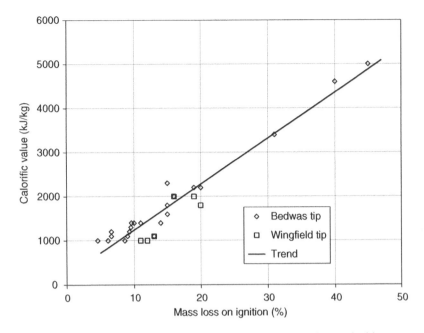

Figure 10.1   Correlation between calorific value and mass loss on ignition.

- material with calorific values <2000 kJ/kg is considered to be unlikely to burn;
- material with calorific value between 2000 kJ/kg and 10,000 kJ/kg should be considered potentially combustible; and
- material with values greater than 10,000 kJ/kg is certainly combustible.

However, a simpler cheaper test, the mass loss on ignition, is sometimes used to assess combustibility. Figure 10.1 shows a correlation between calorific value and mass loss on ignition obtained by the authors, with a suggested trend line of:

$$\text{Calorific value} = 104 \times \text{mass loss on ignition} + 200. \qquad (10.1)$$

The information is obtained from limited data but shows a good correlation between the two properties.

Reference

ICRCL. 1986. *Notes on the Fire Hazards of Contaminated Land*. Inter-Departmental Committee on the Redevelopment of Contaminated Land, Guidance Note 61/84, DETR, UK.

11

Soil-Structure Interfaces

The stability of many foundations and earth structures depends not only on the properties of the founding and surrounding soil but also on the adhesion and friction at the soil-structure interface, while interface friction itself depends on the pressures between the structure and soil. However, while it is normal to measure soil properties, including shear strength, direct measurement of the interaction between structures and the adjacent soil is rarely measured in routine ground investigations.

This chapter gives typical values for properties associated with soil-structure interfaces and, where appropriate, relationships between these values and soil properties.

11.1 Lateral Pressures in a Soil Mass

Consideration of lateral pressures is usually associated with the design of retaining walls, basement walls, pile foundations, pipelines and tunnels, where interest is focused on the minimum and maximum lateral pressures that can occur; that is, on the coefficients of active and passive pressure. Approximate solutions for active and passive pressure problems can be obtained using the simple Coulomb (1776) wedge theory or by consideration of the maximum shear stress at failure (Rankine 1857).

Soil Properties and their Correlations, Second Edition. Michael Carter and Stephen P. Bentley.
© 2016 John Wiley & Sons, Ltd. Published 2016 by John Wiley & Sons, Ltd.

However, while these theories give reasonable values for active pressures, they can give significant underestimates of lateral pressures which can lead to unsafe designs. It is therefore more usual to obtain coefficients of earth pressure using analyses that postulate curved failure surfaces (Caquot and Kerisel 1966; Terzaghi and Peck 1967; Terzaghi et al. 1996). Graphs and equations giving active and passive pressures for various angles of shearing resistance, backfill angle, retaining wall angle and wall friction are available from a number of sources, including Kerisel and Absi (1990).

11.1.1 Earth Pressure at Rest

Active and passive pressures represent the limiting values of lateral earth pressure, when the soil has reached a failure condition, and require a certain amount of movement for pressures to attain these values. The earth pressure at rest is intermediate between these limitations and represents the lateral pressure within a soil mass where no movement has taken place. This can be of practical importance in the calculation of design pressures behind rigid structures, such as strutted retaining walls and heavy gravity retaining walls, in which movement may be insufficient to allow the soil to reach an active state. For such conditions, it is useful to be able to estimate the value of horizontal stress in the undisturbed ground. This cannot be obtained from theoretical considerations of limit equilibrium, as is the case for active and passive pressures, but depends on the geological history of the soil. However, using an approximate theory (Kezdi 1974) the coefficient of earth pressure at rest K_0 for a normally consolidated soil can be related to the angle of shearing resistance φ:

$$K_0 = 1 - \sin\varphi. \tag{11.1}$$

This relationship has been found to hold reasonably true for normally consolidated sands and clays, as indicated in Figures 11.1 and 11.2. In addition, a relationship between K_0 and plasticity index has been obtained by Massarsch (1979), as shown in Figure 11.3. The above relationships are valid for normally consolidated clays but for overconsolidated clays the value of K_0 is heavily dependent on the overconsolidation ratio. For these clays, K_0 can be estimated from Figure 11.4 which shows relationships between K_0 and overconsolidation ratio for clays of different plasticity index values.

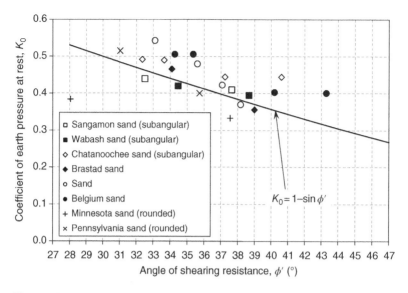

Figure 11.1 Correlation between coefficient of earth pressure at rest and angle of shearing resistance for normally consolidated sands. Adapted from Al-Hussaini and Townsend (1975).

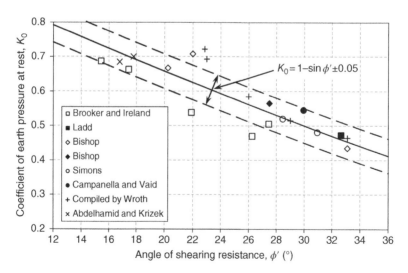

Figure 11.2 Correlation between coefficient of earth pressure at rest and angle of shearing resistance for normally consolidated clays. Adapted from Ladd *et al.* (1977). Data from Brooker and Ireland (1965), Ladd (1975), Bishop (1958), Simons (1958), Campanella and Vaid (1972), Wroth (1984) and Abdelhamid and Krizek (1976).

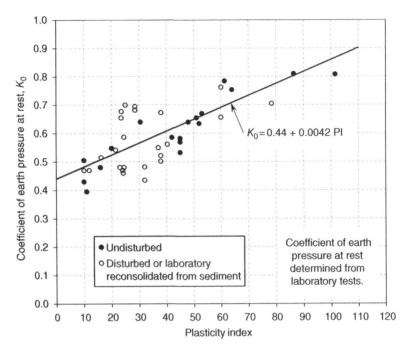

Figure 11.3 Correlations between coefficient of earth pressure at rest and plasticity index. Adapted from Massarsch (1979). Reproduced with permission of the British Geotechnical Association.

11.2 Friction and Adhesion at Interfaces

Values of friction and adhesion between structures and soil are difficult to measure, so are often estimated. Table 11.1 suggests typical values that may be used in preliminary or non-critical analyses (US Navy 1982).

11.2.1 Values Relating to Specific Types of Structure

In addition to the values given in Table 11.1, specific factors are also used for structures such as piles and earth reinforcement. Typical values for some of these factors are discussed below for each type of structure.

11.2.1.1 Piles in Cohesive Soils

As with footings, piles in cohesive soils are usually designed for short-term stability considerations, using undrained shear strength. The adhesion along

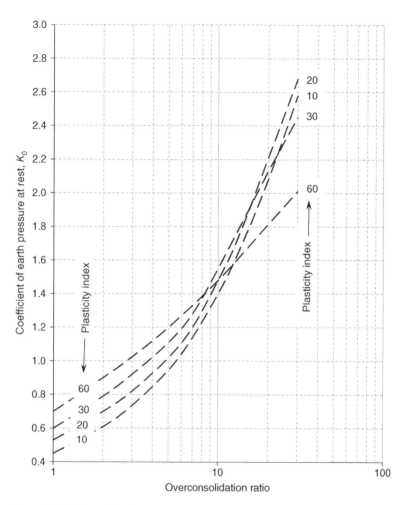

Figure 11.4 Correlation between the coefficient of earth pressure at rest and overconsolidation ratio for clays of various plasticity indices. Adapted from Massarsch (1979). Reproduced with permission of the British Geotechnical Association.

the sides of piles is usually taken as a proportion α of the undrained shear strength of the clay. Typically, a value of $\alpha = 0.45$ is used but this may be reduced in overconsolidated and/or fissured clays: for instance, in London Clay which is both overconsolidated and fissured, a value of 0.3 is often recommended. For heavily overconsolidated, hard clays, α may be reduced to as little as 0.1.

Table 11.1   Typical values of friction and adhesion at interfaces. Adapted from US Naval Publications and Forms Center, US Navy (1982).

Materials	Friction, $\tan \delta$ (°)	Adhesion, c_a (kPa)
Mass concrete or masonry against rocks and soils		
Clean sound rock	0.7	
Clean gravel, gravel-sand mixtures, coarse sand	0.55–0.60	
Clean fine to medium sand, silty medium to coarse sand, silty or clayey gravel	0.45–0.55	
Clean fine sand, silty or clayey fine to medium sand	0.35–0.45	
Fine sandy silt, non-plastic silt	0.30–0.35	
Very stiff and hard residual or overconsolidated clay	0.40–0.50	
Stiff clay and silty clay	0.30–0.35	
Steel sheet piles against soils		
Clean gravel, gravel-sand mixtures, well-graded rock fill	0.40	
Clean sand, silty sand-gravel mixtures, single-size hard rock fill	0.30	
Silty sand, gravel or sand mixed with silt or clay	0.25	
Fine sandy silt, non-plastic silt	0.20	
Soft clay and clayey silt		5–30
Stiff and hard clay and clayey silt		30–60
Formed concrete or concrete sheet piling against soil		
Clean gravel, gravel-sand mixtures, well-graded rock fill	0.40–0.50	
Clean sand, silty sand-gravel mixtures, single-size hard rock fill	0.30–0.40	
Silty sand, gravel or sand mixed with silt or clay	0.30	
Fine sandy silt, non-plastic silt	0.25	
Soft clay and clayey silt		10–35
Stiff and hard clay and clayey silt		35–60
Various structural materials		
Masonry on masonry, and igneous and metamorphic rocks:		
dressed soft rock on dressed soft rock	0.70	
dressed hard rock on dressed soft rock	0.65	
dressed hard rock on dressed hard rock	0.55	
Masonry on wood (cross-grain)	0.50	
Steel on steel at steel-pile interlocks	0.30	

Note: the numbers are approximate values and require sufficient movement for failure to occur. Where friction factor only is shown, the effect of adhesion is included in the friction factor.

11.2.1.2   Piles in Granular Soils

Skin friction values recommended by Broms (1979) for driven steel, concrete and wooden piles in granular soils are given in Table 11.2. These values may also be used as a guide for bored piles but the relative density should always be assumed to be loose (φ <35°), regardless of the natural condition of the soil, to allow for loosening of the soil during boring.

Since frictional forces depend on the applied normal pressure between surfaces, a further consideration for piles in granular soils is the lateral earth pressure on the pile. For this, Broms (1979) recommends taking a proportion K_s of the effective overburden pressure, as indicated in Table 11.2. Note that the values given in the table apply to driven piles which compress the surrounding soil; hence the K_s factors are greater than 1 for dense deposits (φ >35°). Again, this is not the case for bored piles where loose conditions, with φ <35°, should be assumed regardless of the soil's natural relative density, and values of K_s should normally be assumed to be 1 or less unless there is specific reason to believe that the soil will have been compacted by, for instance, cast-*in-situ* concrete being compacted into the hole.

11.2.1.3   Soil Reinforcement and Soil Nails

For geogrids and geotextiles, friction between the grid and the soil is based on a proportion α of the angle of shearing resistance, so that the skin friction per unit area is given by:

$$\text{skin friction per unit area} = 2\alpha\sigma_v' \tan\varphi, \qquad (11.2)$$

where the factor 2 takes account of the top and bottom faces of the reinforcement and σ_v' is the effective vertical earth pressure on the reinforcement.

Table 11.2   Values of K_s and δ for driven piles. Adapted from Broms (1979).

Pile material	Skin friction angle, δ	Earth pressure coefficient, K_s	
		Low relative density (ϕ <35°)	High relative density (ϕ >35°)
Steel	20°	0.5	1.0
Concrete	3φ/4	1.0	2.0
Wood	2φ/3	1.5	4.0

The value of α depends on the individual material and should be supplied by the manufacturer, but a value of 0.95 is typically quoted for geogrids, with 0.8 for geotextiles. As described in Chapter 6, it is normally considered appropriate to use constant-volume shear strength rather than peak strength, because of the movement that may take place with flexible reinforcement even during normal service conditions. This is often estimated by applying a reduction factor of typically 20% to the peak strength value. However, care should be taken to check whether the factor α for the design method being used already includes this reduction, to avoid an over-conservative design.

A similar situation arises with soil nails, which are grouted into the surrounding soil. For these, a design value of $\alpha = 0.9$ is typical. However, since the earth pressure on a soil nail acts all around, not just on the top and bottom as for a geogrid, earth pressure is weighted between vertical and horizontal. The effective earth pressure σ_n acting all around the nail is typically taken as:

$$\sigma_n = \frac{1}{4}\left(3 + K_0\right)\sigma_v. \qquad (11.3)$$

11.2.1.4 Retaining Walls

For soil-structure friction beneath the base and along the sides of retaining walls, Table 11.1 values, given by the US Navy (1982) may be used, at least for a preliminary assessment. However, many gravity walls incorporate a base that extends behind the wall and this base, and the soil above it, are considered to be a part of the wall for stability calculations. This means that the back face of the wall is effectively a soil-soil interaction, so it may be assumed that the angle of shearing resistance is appropriate as the effective rear face friction on the wall. However, it is usually recommended that, because of the distribution of stresses behind gravity walls, the design rear wall friction value should be limited to $^2/_3\varphi$.

References

Abdelhamid, M.S. and Krizek, R. J. 1976. At rest lateral earth pressure of a consolidating clay. *Proceedings of ASCE Journal of the Geotechnical Engineering Division* 102: 721–738.

Al-Hussaini, M. M. and Townsend, F. C. 1975. Investigation of K0 testing in cohesionless soils. Technical Report S-75-11, US Army Engineer Waterways Experimental Station, Vicksburg, Mississippi, 70 pp.

Bishop, A. W. 1958. Test requirements of measuring the coefficient of earth pressure at rest. *Proceedings of Conference on Earth Pressure Problems*, Brussels, I: 2–14.

Broms, B. 1979. Negative skin friction. *Proceedings of the 6th Asian regional Conference on Soil Mechanics and Foundation Engineering*, Singapore, 2: 41–76.

Brooker, E. W. and Ireland, H. O. 1965. Earth pressures at rest related to stress history. *Canadian Geotechnical Journal* 2: 1–15.

Campanella, R. G. and Vaid, Y. P. 1972. A simple K0-triaxial cell. *Canadian Geotechnical Journal* 9: 249–260.

Caquot, A. and Kerisel, J. 1966. *Traite de Mechanique des Sols.* Gauthier-Villars, Paris, 509 pp.

Coulomb, C. A. 1776. Essai sur une application des règles de maximis et minimis à quelques problèmes de statique, relatifs à l'architecture. *Memoires de l'Academie Royale des Sciences, Paris*, 3, p38.

Kerisel, J. and Absi, E. 1990. *Earth Pressure Tables.* Balkema, Rotterdam.

Kezdi, A. 1974. *Handbook of Soil Mechanics.* Vol. 1: Soil Physics. Elsevier, Philadelphia.

Ladd, C. C. 1975. *Foundation Design of Embankments Constructed on Connecticut Valley Varved Clays.* Research Report R75-7, Geotechnical Publication 343, Department of Civil Engineering, Massachusetts Institute of Technology, 438 pp.

Ladd, C. C., Foote, R., Ishihara, K., Schlosser, F. and Poulos, H. G. 1977. Stress-deformation and strength characteristics. *Proceedings of 9th International Conference on Soil Mechanics and Foundation Engineering*, Tokyo, 2: 421–494.

Massarsch, K. R. 1979. Lateral earth pressure in normally consolidated clay. *Proceedings of 7th European Conference on Soil Mechanics and Foundation Engineering*, Brighton, 2: 245–250.

Rankine, W. J. M. 1857. On the stability of loose earth. *Philosophical Transactions of the Royal Society, London*, 147: 9–27.

Simons, N. E. 1958. Discussion of test requirements for measuring the coefficient of earth pressure at rest. *Proceedings of Conference on Earth Pressure Problems*, Brussels, 3: 40–53.

Terzaghi, K. and Peck, R. B. 1967. *Soil Mechanics in Engineering Practice.* John Wiley, London, 729 pp.

Terzaghi, K., Peck, R. B. and Mesri, G. 1996. *Soil Mechanics in Engineering Practice*, 3rd edition. John Wiley, New York, USA.

US Navy. 1982. *Design Manual: Soil Mechanics, Foundations and Earth Structures.* Navy Facilities Engineering Command, Navfac, US Naval Publications and Forms Center.

Wroth, C. P. 1984. The interpretation of insitu soil tests. *Geotechnique* 34: 449–489.

Appendix A

Soil Classification Systems

Many typical soil properties given in this book refer to soil classes, as defined in commonly used soil classification systems. This appendix gives a summary of the more common systems and the definitions of the soil classes within each system.

The purpose of a soil classification system is to group together soils with similar properties or attributes. From the engineering standpoint, it is the geotechnical properties such as permeability, shear strength and compressibility that are important.

The first step in classifying a soil is to identify it. To be of practical value, a classification system should permit identification by either inspection or testing, and tests should be as simple as possible. In this respect, tests that require disturbed samples are preferable: not only do they dispense with the need for undisturbed sampling or field testing but, in addition, the properties they measure do not depend on the structure of the soil mass. Properties such as grain size, mineral composition, organic matter content and soil plasticity are therefore preferred as a basis for a classification system rather than properties such as moisture content, density and shear strength.

Implicit in the concept that soils with similar properties can be grouped together is the assumption that correlations exist between the various soil

Soil Properties and their Correlations, Second Edition. Michael Carter and Stephen P. Bentley.
© 2016 John Wiley & Sons, Ltd. Published 2016 by John Wiley & Sons, Ltd.

properties. That this is true is borne out not only by the success of soil classification systems but also by the many correlations given throughout this text. However, since correlations are only approximate, classification systems can give only a rough guide to suitability and behaviour: a limitation which must be appreciated if classification systems are to be used sensibly. This is particularly important where a classification system based on the testing of disturbed samples is used to predict properties that depend on the state of the soil mass. For instance, since the shear strength of a clay is heavily influenced by factors such as moisture content and field density, a classification system based on soil plasticity tests alone cannot be expected to predict bearing capacity to any great accuracy. In this respect, classification systems are more applicable where soils are used in remoulded form than where they are used in their natural state, and it is not surprising that the most commonly used engineering soil classification systems were all developed for earthworks, highways and airports work.

A.1   Systems Based on the Casagrande System

A.1.1   The Unified System

The Unified system is the oldest system to be widely adopted, and variations of this system still represent probably the most widely used form of soil classification in the English-speaking world. It was developed from a system proposed by Casagrande (1948) and referred to as the Airfield Classification System. Coarse-grained soils (sands and gravels) are classified according to their grading, and fine-grained soils (silts and clays) and organic soils are classified according to their plasticity, as indicated in Table A1 which shows the basic classifications.

An ingenious feature of the system is the differentiation of silts and clays by means of the plasticity chart shown in Figure A1. The position of the A-line was fixed by Casagrande based on empirical data. However, the use of this chart for distinguishing silt from clay leads to a subtle change in the definition of silt compared with its definition in terms of particle size – for instance, it is defined as comprising particles of 5-75 μm in the AASHTO system and 2-63 μm in the BS system. The BS system suggests the use of the term 'M-soil' for silt defined in terms of its plot on the plasticity chart, to avoid confusion with silt defined in terms of particle size, but the term has never gained popular acceptance.

Table A1 The Unified soil classification system.

Major divisions			Typical names	Group symbols
COARSE GRAINED SOILS (More than half the material is larger than 75μm)	GRAVELS (More than half of coarse fraction is larger than 4.75mm)	Clean gravels (little or no fines)	Well graded gravels, gravel-sand mixtures, little or no fines	GW
			Poorly graded gravels, gravel-sand mixtures, little or no fines	GP
		Gravels with fines (appreciable amount of fines)	Silty gravels, poorly graded gravel-sand-silt mixtures	GM
			Clayey gravels, poorly graded gravel-sand-clay mixtures	GC
	SANDS (More than half of coarse fraction is smaller than 4.75mm)	Clean sands (little or no fines)	Well graded sands gravelly sands, little or no fines	SW
			Poorly graded sands gravelly sands, little or no fines	SP
		Sands with fines (appreciable amount of fines)	Silty sands poorly graded sand-silt mixtures	SM
			Clayey sands poorly graded sand-clay mixtures	SC
FINE GRAINED SOILS (More than half the material is smaller than 75μm)	SILTS AND CLAYS - liquid limit less than 50		Inorganic silts and very fine sands, rock flour, silty or clayey fine sands with slight plasticity	ML
			Inorganic clays of low to medium plasticity, gravelly clays. sandy clays, silty clays, lean clays	CL
			Organic silts and organic silt-clays of low plasticity	OL
	SILTS AND CLAYS - liquid limit greater than 50		Inorganic silts, micaceous or diatomaceous fine sandy or silty soils, elastic silts	MH
			Inorganic clays of high plasticity, fat clays	CH
			Organic clays of medium to high plasticity	OH
Highly organic soils			Peat and other highly organic soils	Pt

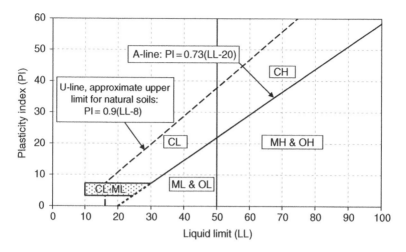

Figure A1   Plasticity chart for the Unified/ASTM soil classification system.

A.1.2   The ASTM System

An advantage of the Unified system is that it can easily be extended to include more soil groups, giving a finer degree of classification if required. The American Association for Testing and Materials (ASTM 2006) has adopted the Unified system as a basis for the ASTM soil classification, entitled *Standard Test Method for Classification of Soils for Engineering Purposes*, designation D2487, which uses an extended classification over that given in Table A1, and gives more standardisation by defining the classes in terms of laboratory test values. A summary of the ASTM classes is given in Table A2 for coarse-grained soils and in Table A3 for fine-grained soils.

A.1.3   The British Standard System

The British Standard classification system (BS 1990) is, like the Unified system, also based on the Casagrande classification but the definitions of sand and gravel are slightly different, to be in keeping with other British Standards, and the fine-grained soils are divided into five plasticity ranges (Fig. A2) rather than the simple 'low' and 'high' divisions of the Unified and the original Casagrande systems. In addition, a considerable number of sub-groups have been introduced. The definitions of the soil groups and sub-groups are given in Table A4.

Table A2 The Unified soil classification system. Extended soil groupings for coarse-grained soils defined by specific laboratory test values, as used in ASTM D2487.

Major divisions			Test criteria		Group symbols
			Passing 75μm sieve	Other test criteria*	
COARSE GRAINED SOILS (More than half the material is larger than 75μm)	GRAVELS (More than half of coarse fraction is larger than 4.75mm)	Clean gravels (little or no fines)	< 5%	$C_u = \dfrac{D_{60}}{D_{10}}$ greater than or equal to 4; $C_c = \dfrac{D_{30}^{2}}{D_{10} \times D_{60}}$ between 1 and 3	GW
				Not meeting both criteria for GW	GP
		Gravels with fines (appreciable amount of fines)	> 12%	Plots below the A-line or PI < 4	GM
				Plots above the A-line and PI > 7	GC
				Plots in hatched area of plasticity chart, marked CL-ML	GC-GM
		Gravels with intermediate amount of fines	5-12%	Meets the criteria in this column for GW and GM	GW-GM
				Meets the criteria in this column for GW and GC	GW-GC
				Meets the criteria in this column for GP and GM	GP-GM
				Meets the criteria in this column for GP and GC	GP-GC
	SANDS (More than half of coarse fraction is smaller than 4.75mm)	Clean sands (little or no fines)	< 5%	$C_u = \dfrac{D_{60}}{D_{10}}$ greater than or equal to 6; $C_c = \dfrac{D_{30}^{2}}{D_{10} \times D_{60}}$ between 1 and 3	SW
				Not meeting both criteria for SW	SP
		Sands with fines (appreciable amount of fines)	> 12%	Plots below the A-line or PI < 4	SM
				Plots above the A-line and PI > 7	SC
				Plots in hatched area of plasticity chart, marked CL-ML	SC-SM
		Sands with intermediate amount of fines	5–12%	Meets the criteria in this column for SW and SM	SW-SM
				Meets the criteria in this column for SW and SC	SW-SC
				Meets the criteria in this column for SP and SM	SP-SM
				Meets the criteria in this column for SP and SC	SP-SC

* C_u is the coefficient of uniformity and C_c is the coefficient of curvature.

Table A3 The Unified soil classification system. Extended soil groupings for fine-grained soils defined by specific laboratory test values as used in ASTM D2487.

Major divisions		Test criteria			Group symbols
		Liquid limit	Plasticity index	Other test criteria	
FINE GRAINED SOILS (More than half the material is smaller than 75μm)	SILTS AND CLAYS – liquid limit less than 50	< 50	< 4	Inorganic. or plots below the A-line	ML
		< 50	> 7	Inorganic. Plots on or above the A-line	CL
		< 50	–	Organic. $(LL_{oven\ dried})/(LL_{not\ dried}) < 0.75$	OL
	SILTS AND CLAYS – liquid limit greater than 50	≥ 50	–	Inorganic. Plots below the A-line	MH
		≥ 50	–	Inorganic. Plots on or above the A-line	CH
		≥ 50	–	Organic. $(LL_{oven\ dried})/(LL_{not\ dried}) < 0.75$	OH
		≥ 50	–	Inorganic. Plots in hatched area of plasticity chart	CL-ML
Highly organic soils		–	–	Peat, muck and other highly organic soils	Pt

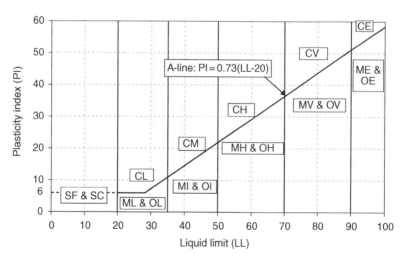

Figure A2 Plasticity chart for the British Standard soil classification system.

Table A4 The British Standard soil classification system.

Soil Groups		Description and Identification		Sub-Groups		Liquid Limit	Fines (% <0.06 mm)
Slightly silty or clayey GRAVEL	G	Well graded GRAVEL	GW	–	–		0–5
		Poorly graded GRAVEL	GP	Uniformly graded*	GPu		
				Gap graded	GPg		
Silty or clayey GRAVEL	G-F	Silty GRAVEL	G-M	Well graded	GWM		5–15
				Poorly graded	GPM		
		Clayey GRAVEL	G-C	Well graded	GWC		
				Poorly graded	GPC		
Very silty or clayey GRAVEL	GF	Very silty GRAVEL	GM	Subdivisions as for GC	GML etc.	As GC	15–35
		Very clayey GRAVEL	GC	Low plasticity fines	GCL	<35	
				Intermediate plasticity fines	GCI	35–50	
				High plasticity fines	GCH	50–70	
				Very high plasticity fines	GCV	70–90	
				Extremely high plasticity fines	GCE	>90	
Slightly silty or clayey SAND	S	Well graded SAND	SW	–	–		0–5
		Poorly graded SAND	SP	Uniformly graded*	SPu		
				Gap graded	SPg		
Silty or clayey SAND	S-F	Silty SAND	S-M	Well graded	SWM		5–15
				Poorly graded	SPM		
		Clayey SAND	S-C	Well graded	SWC		
				Poorly graded	SPC		
Very silty or clayey SAND	SF	Very silty SAND	SM	Subdivisions as for SC	SML etc.	As SC	15–35
		Very clayey SAND	SC	Low plasticity fines	SCL	<35	
				Intermediate plasticity fines	SCI	35–50	
				High plasticity fines	SCH	50–70	
				Very high plasticity fines	SCV	70–90	
				Extremely high plasticity fines	SCE	>90	

GRAVELS (>50% of coarse material is of gravel size – >2mm)

SANDS (>50% of coarse material is of sand size – 0.06mm to 2.00mm)

COARSE SOILS (<35% fines)

Major group	Subgroup	Code	Soil	Symbol	Plasticity	Symbol	Range
FINE SOILS (>35% fines)							
Gravelly or sandy SILTS or CLAYS (35–65% fines)	Gravelly SILT or gravelly CLAY*	FG	Gravelly* SILT	MG	As for CG	MLG etc.	As CG
			Gravelly* CLAY	CG	Low plasticity	CLG	<35
					Intermediate plasticity	CIG	35–50
					High plasticity	CHG	50–70
					Very high plasticity	CVG	70–90
					Extremely high plasticity	CEG	>90
	Sandy SILT or sandy CLAY*	FS	Sandy* SILT	MS	As for CG	MLS etc.	As CG
			Sandy* CLAY	CS	As for CG	CLS etc.	As CG
SILTS and CLAYS (>65% fines)	SILT or CLAY	F	SILT	M	As for C	ML etc.	As C
			CLAY	C	Low plasticity	CL	<35
					Intermediate plasticity	CI	35–50
					High plasticity	CH	50–70
					Very high plasticity	CV	70–90
					Extremely high plasticity	CE	>90
ORGANIC SOILS			Letter 'O' suffixed to any group or subgroup symbol. e.g. MHO – organic silt of high plasticity.				
PEAT		Pt	Peat soils consist predominantly of plant remains (fibrous or amorphous)				

* Gravelly if >50% coarse material is gravel sized: sandy if >50% coarse material is sand sized.

^ Material is generally considered to be uniformly graded if it has a uniformity coefficient of less than 6, where the uniformity coefficient is defined as $U = D_{60}/D_{10}$ where D_{60} and D_{10} are the 60% and 10% particle sizes, respectively.

A.2   The AASHTO System

Unified system and its derivatives classify soil by type rather than by engineering suitability for specific uses, although they can nevertheless be used to infer suitability. By contrast, the system defined by the American Association of State Highway and Transportation Officials (AASHTO 2012) does not classify soils by type (e.g. sands, clays) but simply divides them into seven major groups, essentially classifying soils according to their suitability as subgrades. Some groups may be further divided into subgroups, as indicated in Table A5 which describes the materials present in each group. The definition of each group, based on laboratory testing, is given in Table A6.

As a further refinement in assessing the suitability of materials, each material can be given a 'group index' value defined as:

$$\text{group index} = (F - 35)\left[0.2 + 0.005(LL - 40)\right] + (F - 15)(PI - 10)$$

where F is the percentage passing the 75 μm (0.075 mm) sieve, expressed
as a whole number. This percentage is based only on the material
passing the 75 mm sieve.
LL is the liquid limit and
PI is the plasticity index.

This is usually shown in brackets after the soil class. When applying the formula, the following rules are used:

- the group index is reported to the nearest whole number and, if it is negative, it is reported as zero;
- when calculating the group index of subgroups A-2-6 and A-2-7, only the plasticity index portion of the formula should be used; and
- because of the criteria that define subgroups A-1-a, A-1-b, A-2-4, A-2-5 and group A3, their group index will always be zero, so the group index is usually omitted from these classes.

Originally the group index was used directly to obtain pavement thickness designs, using the 'group index method' but this approach has long since been superseded and the group index is used only as a guide.

Table A5 Description of soil types in the AASHTO soil classification system.

Classification of materials in the various groups applies only to the fraction passing the 75 mm sieve. The proportions of boulder- and cobble-sized particles should be recorded separately and any specification regarding the use or A-1, A-2 or A-3 materials in construction should state whether boulders are permitted.

Granular materials	Silty clay materials

Granular materials

Group A-1. Typically a well graded mixture or stone fragments or gravel, coarse to fine sand and a non-plastic or feebly plastic soil binder. However, this group also includes stone fragments, gravel, coarse sand, volcanic cinders, etc. without soil binder.

Subgroup A-1-a is predominantly stone fragments or gravel, with or without binder.

Subgroup A-1-b is predominantly coarse sand with or without binder.

Group A-3. Typically fine beach sand or desert sand without silty or clayey fines or with a very small proportion or non-plastic silt. The group also includes stream-deposited mixtures of poorly graded fine sand with limited amounts of coarse sand and gravel.

Group A-2. Includes a wide variety of 'granular' materials which are borderline between the granular A-1 and A-3 groups and the silty clay materials of groups A-4 to A-7. It includes all materials with not more than 35% fines which are too plastic or have too many fines to be classified as A-1 or A-3.

Subgroups A-2-4 and A-2-5 include various granular materials whose finer particles (0.425 mm down) have the characteristics of the A-4 and A-5 groups, respectively.

Subgroups A-2-6 and A-2-7 are similar to those described above but whose finer particles have the characteristics of A-6 and A-7 groups, respectively.

Silty clay materials

Group A-4. Typically a non-plastic or moderately plastic silty soil usually with a high percentage passing the 0.075 mm sieve. The group also includes mixtures of silty fine sands and silty gravelly sands.

Group A-5. Similar to material described under group A-4 except that it is usually diatomaceous or micaceous and may be elastic as indicated by the high liquid limit.

Group A-6. Typically a plastic clay soil having a high percentage passing the 0.075 mm sieve. Also mixtures of clayey soil with sand and fine gravel. Materials in this group have a high volume change between wet and dry states.

Group A-7. Similar to material described under group A-6 except that it has the high liquid limit characteristic of group A-5 and may be elastic as well as subject to high volume change.

Subgroup A-7-5 materials have moderate plasticity indices in relation to the liquid limits and may be highly elastic as well as subject to volume change.

Subgroup A-7-6 materials have high plasticity indices in relation to the liquid limits and are subject to extremely high volume change.

Group A-8. Includes highly organic materials. Classification of these materials is based on visual inspection and is not related to grading or plasticity.

Table A6 The AASHTO soil classification system.

GENERAL CLASSIFICATION	GRANULAR MATERIALS (35% or less passing 0.075mm sieve)							SILT-CLAY MATERIALS (> 35% passing 0.075mm sieve)				
	A-1		A-3	A-2				A-4	A-5	A-6	A-7	
Group classification	A-1-a	A-1-b		A-2-4	A-2-5	A-2-6	A-2-7				A-7-5	A-7-6
Sieve analysis, % passing:												
2mm	50 max	–	–	–	–	–	–	–	–	–	–	–
0.425mm	30 max	50 max	51 min	–	–	–	–	–	–	–	–	–
0.075mm	10 max	25 max	10 max	35 max	35 max	35 max	35 max	36 min	36 min	36 min	36 min	36 min
Fraction passing 0.425mm:												
Liquid limit	–		–	40 max	41 min	40 max	41 min	40 max	41 min	40 max	41 min	41 min
Plasticity index	6 max		Non-plastic	10 max	10 max	11 min	11 min	10 max	10 max	11 min	11 min*	11 min*
Usual types of significant constituents	Stone fragments, gravel, sand		Fine sand	Silty or clayey gravel and sand				Silty soils		Clayey soils		
General rating as a subgrade	Excellent to good							Fair to poor				

* Plasticity index of A-7-5 subgroup is equal to or less than liquid limit − 30.
* Plasticity index of A-7-6 subgroup is greater than liquid limit − 30.

Table A7 Comparison of soil groups in the Unified and British Standard soil classification systems.

British Standard system			Unified/ASTM system, most probable group
Group	Subgroup	Subdivision	
G	GW	GPu	GP
	GP	Gpg	GP
G-F	GM	GWM	GW-GM
		GPM	GP-GM
	GC	GWC	GW-GC
		GPC	GP-GC
GF	GM		GM
	GC		GC
S	SW		SW
	SP	SPu	SP
		SPg	SP
S-F	S-M	SWM	SW-SM
		SPM	SP-SM
	S-C	SWC	SW-SC
		SPC	SP-SC
SF	SM		SM
	SC		SC
FG	MG	MLG, MIG MHG, MVG	ML,OL
		MEG	MH, OH
	CG	CLG, CIG CHG, CVG	CL
		CEG	CH
FS	MS	MLS, MIS MHS,MVS	ML, OL
		MES	MH, OH
	CS	CLS, CIS	CL
		CHS,CVS, CES	CH
F	M	ML,MI MH,MV,ME	ML, OL MH, OH
	C	CL, CI CH, CV, CE	CL CH
Pt			Pt

A.3   Comparison of the Unified, AASHTO and BS Systems

A correlation between the BS and Unified/ASTM systems is given in Table A7. Because the two systems share a common origin, it is possible to correlate the soil groups with a reasonable degree of confidence. However, minor differences between the systems mean that the possibility of ambiguity can arise, so that equivalence of the classes shown in the table may not hold true for all soils.

The totally different basis of the AASHTO system means that there is no direct equivalence between it and the groups of the Unified system but probable equivalent classes, applicable for most soils, are indicated in Table A8. A full comparison of the Unified, AASHTO and now-superseded US Federal Aviation Agency (FAA) systems is given by Liu (1967).

Table A8   Comparison of soil groups in the Unified and AASHTO soil classification systems.

Comparing Unified ASTM with AASHTO		Comparing AASHTO with Unified ASTM	
Unified/ASTM soil group	Most probable AASHTO soil group	AASHTO soil group	Most probable Unified/ASTM soil group
GW	A-1-a	A-1-a	GW, GP
GP	A-1-a	A-1-b	SW, SP, GM, SM
GM	A-1-b, A-2-4 A-2-5, A-2-7	A-3	SP
GC	A-2-6, A-2-7	A-2-4	GM, SM
SW	A-1-b	A-2-5	GM, SM
SP	A-3, A-1-b	A-2-6	GC, SC
SM	A-1-b, A-2-4 A-2-5, A-2-7	A-2-7	GM, GC, SM, SC
SC	A-2-6, A-2-7	A-4	ML, OL
ML	A-4, A-5	A-5	OH, MH, ML, OL
CL	A-6, A-7-6	A-6	OL
OL	A-4, A-5	A-7-5	OH, OM
MH	A-7-5, A-5	A-7-6	CH, CL
CH	A-7-6		
OH	A-7-5, A-5		
Pt	–		

References

AASHTO. 2012. *Classification of soils and soil-aggregate mixtures for highway construction purposes*. M145-91. American Association of State Highway and Transportation Officials. Washington, DC, USA.

ASTM. 2006. *Standard practice for classification of soils for engineering purposes*. Unified Soil Classification System). D2487. West Conshohocken, PA, USA.

BS. 1990. *Methods of test for soils for civil engineering purposes*. British Standards Institution, BS 1377.

Casagrande, A. 1948. Classification and identification of soils. *Transactions ASCE* 113: 901–932.

Liu, T. K. 1967. A review of engineering soil classification systems. *Highway Research Board Record* 156: 1–22.

Appendix B

Sampling Methods

Soil sampling methods are clearly related to testing requirements: soils to be left in their natural state, such as beneath foundations, within natural and cut slopes and beneath fills require undisturbed samples for testing to reflect their *in situ* properties; soils to be excavated and compacted in a specified way, such as for earthworks including highway and railway fills, generally require only disturbed samples to assess their relevant properties. However, a further consideration, that is often less well appreciated, is that the type of sampling can significantly affect the measured properties of soil. This appendix summarises the most common forms of sampling and, where appropriate, comments on the relevance to the quality of information obtained when the samples are tested.

In addition to the sampling methods used, a further factor that can have a significant effect on sample quality, and therefore on the accuracy of measured soil properties, is the skill and conscientiousness of the personnel. In this respect, the trend over recent years of reducing site supervision on ground investigations is to be regretted since the interests of drillers, whose pay, in a competitive industry, depends partly or even largely on progress made, are not always aligned with the interests of the geotechnical engineers whose need is for reliable test data. These aspects are included in the discussions below, where appropriate.

Soil Properties and their Correlations, Second Edition. Michael Carter and Stephen P. Bentley.
© 2016 John Wiley & Sons, Ltd. Published 2016 by John Wiley & Sons, Ltd.

B.1   Pits and Borings

The first stage in obtaining samples is to excavate a test hole of some kind so that samples can be obtained at the required depths. A wide variety of methods are used, some of the more common of which are described in the following sections.

B.1.1   Trial Pits

Not only do trial pits provide the cheapest method of forming exploratory holes, but they also give the clearest picture of the ground condition. Their main limitation is that they can be dug to only a limited depth; typically pits are limited to about 3 m depth, although they may be dug to 5 m or more with the right excavator and in suitable ground conditions. Pits are almost always dug by excavator because of speed and cost considerations, but may be dug by hand where access prohibits the use of mechanical equipment or where labour costs are very low.

A further consideration is stability, and shoring may be required in loose or soft ground. Excavations are likely to be unstable even with shoring in sands and silts below the water table. Entering a trial pit always carries a risk and should avoided where there is any risk of instability, especially for pits more than 1.2 m deep which should not normally be entered.

Disturbed samples are usually taken from the bucket of the excavator, or from the side of a shallow pit. Undisturbed samples may be obtained by hand-cutting blocks of soil in shallow pits, but it is more common to use sample tubes (described in Section B.2.2) to obtain standard undisturbed samples. These can be pushed into the bottom of a pit by fixing a standard adaptor (see Fig. B3) on the top to protect it, and pushing it in using the excavator bucket. A block of wood is placed on top of the adapter, between it and the bucket or hammer, to protect the top of the adapter. With a good operator, pushing samples in with an excavator bucket often produces samples that are less disturbed than using a standard driving hammer in a borehole. For hand dug pits, a sledge hammer can be used to drive the tube. The pit sides can also be photographed to give a clear visual record of the ground conditions.

Pits may also be filled with water and the level monitored with time to assess the soil permeability or the ground suitability for soakaway drainage systems.

B.1.2   Light Cable Percussion Borings

This basic form of boring dates back well over half a century and may appear to be rather old-fashioned. It is also slow and expensive compared with more modern alternatives. Set against this, it has the advantage of being able to penetrate nearly all soil conditions including granular and cohesive soils, obstructions such as boulders and buried objects, and can progress holes below the water table. This versatility has ensured that the method is still frequently used. An example of a percussion boring rig, along with some of the equipment commonly used with it, is shown in Figure B1.

Boreholes are usually 150 mm or 200 mm in diameter and steel casing is generally required throughout most or all of the borehole depth. For deeper holes, 200 mm casing may be used with 150 mm casing inserted inside it when boring becomes difficult. Progress is achieved by repeatedly dropping an auger, consisting of an open steel tube, to the bottom of the borehole. Once brought back to the surface, the clay held inside it is removed with a crowbar, utilising the slots at the sides (see Fig. B1). Holes may be up to 30 m deep unless difficult drilling conditions are encountered, and can be as deep as 50 m in particularly good drilling conditions, but this is unusual and drilling at this depth is very slow. In sands and gravels a valve is fitted to the lower end of the auger to trap material entering it; this is the 'shell' or 'sand auger'. A variety of specialist tools may be used for specific conditions, most commonly a heavy chisel used to break up obstructions.

The casing is pushed down as drilling proceeds, normally by hammering it down using 'sinker bars' (see Fig. B1) acting against an iron bar threaded across the top of the casing. Methods vary, however, and some drillers prefer to surge the casing, repeatedly lifting it and allowing it to fall back into the ground. This can be satisfactory in some clays, but may cause disturbance problems in some ground conditions.

Disturbed samples can be retrieved from the auger or sand auger in any soils, and undisturbed samples can be obtained in cohesive soils using a simple open-tube sampler or, in soft ground, a piston sampler (see Sections B.2.2 and B.2.3 below). Vane testing, to obtain the undrained shear strength, may also be carried out at the bottom of the hole in soft to firm clays, but this is fairly unusual. Undisturbed sampling is not possible in granular soils, and *in situ* ground conditions are normally estimated from the results of standard penetration tests.

Water is often added in conjunction with the sand auger to help liquefy sand at the bottom of the borehole and encourage it to enter the sand auger,

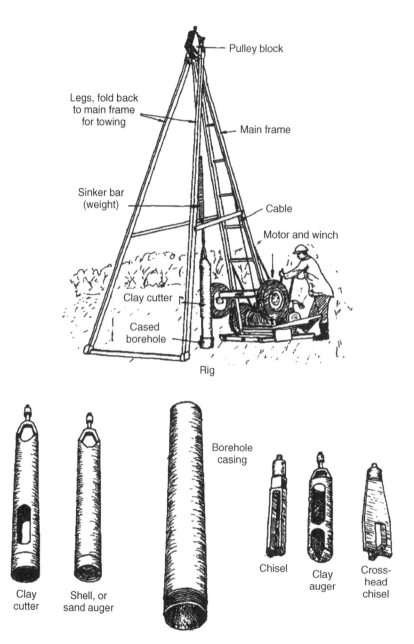

Figure B1 Light cable percussive rig and tools.

to progress the hole. This disturbed material at the bottom of the hole must be cleaned out before a standard penetration test is carried out, or the N-value obtained will be lower than it should be for the *in situ* relative density.

In sands below the water table, water flowing into the bottom of the borehole can drag the surrounding sand with it so that the hole is not progressed even though material is being extracted. This not only inhibits progress but also results in a zone of reduced relative density around the bottom of the hole, again resulting in unrealistically low SPT N-values. The correct procedure to overcome this problem is to keep the hole surcharged with water above the water table, so that inflows are stopped. However, this usually requires the use of a water bowser, and may take large amounts of water in more permeable soils. Drillers find it much easier to drive the casing ahead of the borehole as a means of reducing inflows. Unfortunately, this compacts the soil around the bottom of the borehole, which can result in unrealistically high SPT N-values. Thus, in gravels and especially sands, the SPT N-values obtained are critically dependent on the skill and care of the driller and may not reflect true ground conditions.

B.1.3 Rotary Boring

Rotary drilling is normally associated with rock drilling but rotary boring rigs, usually using augers, are increasingly used in soils as an alternative to traditional percussion boring. Rotary boring rigs come in a wide variety of types and sizes; a typical light rig is illustrated in Figure B2. With simple equipment, only disturbed samples are obtained, from arisings taken from the auger, but with some rigs undisturbed sampling is possible; sample tubes being pushed, hammered or drilled into the base of the borehole. While most rigs use an auger, some use coring bits similar to those used for rocks to obtain undisturbed samples during drilling.

Rotary boring provides a faster, cheaper alternative to percussion boring but is more limited in the type of ground conditions it can deal with. Obstructions will usually stop the auger, and holes are not normally lined, so progress through sand and gravel is limited, and impossible below the water table.

B.1.4 Window Samplers and Windowless Samplers

Window sampling was originally introduced as a method of obtaining samples in clays using portable equipment, often light enough to be carried to site where machine access was not possible. A tube is driven into the ground

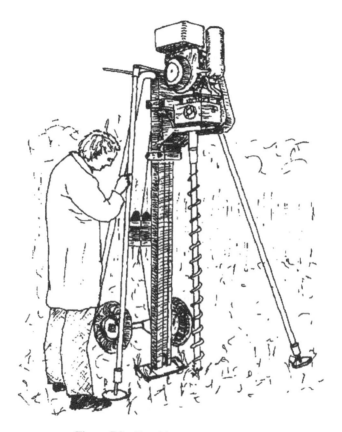

Figure B2   Portable rotary power auger.

using a standard drop hammer, or sometimes a vibrating hammer. This is
then withdrawn and a sample of the soil that has been pushed into the bottom of the tube can be obtained, aided by the tube having two wide slots or
'windows' in the sides. Only disturbed samples are possible, but the *in situ*
shear strength can be obtained by pushing a hand penetrometer or hand vane
into the soil exposed by the window while it is still in the sampler.

After sampling, the sampler is cleaned out using a crowbar and pushed
back in the hole. Further sections of tube are added to extend the depth.
Sample tubes are normally 1 m in length. Because of the lightness of the
equipment, the depth that can be achieved is limited but may be extended by
using progressively smaller tubes, each successive tube being driven inside
the previous one. Even with this method, however, depths are normally

restricted to about 3 m in clays with little or no penetration into sand or gravel or even some gravelly clays.

A development of the window sampler is the powered window sampler which is similar but uses heavier, powered equipment. A further variation, the windowless sampler, allows undisturbed samples to be obtained by keeping the soil in the sample tube, which dispenses with the windows.

Because of its speed and relatively low cost, window and windowless sampling has replaced traditional percussive boring in many instances but is less versatile, being restricted in depth and to ground that will remain open when the sampler is removed, thus precluding its use in granular soils and soft clays. Also, the sample tube diameters are small, typically 35-80 mm, which restricts the testing that can be carried out on samples.

B.2   Sampling and Samplers

B.2.1   Disturbed Samples

Disturbed samples are useful both for identifying and describing the soil and for tests that do not require the soil to be in its natural condition, such as classification tests and any test on soil to be used in earthworks where specimens will be compacted under specific conditions before testing. In sands and gravels, where it is normally impossible to obtain undisturbed material, all samples will be disturbed.

For clay soils with no material above sand size, a 500 g sample is normally sufficient, retained in a jar, small polythene bag or other suitable container to maintain the moisture content. For soils with gravel-sized particles and above, samples will need to be at least 5 kg and may be as large as 50 kg, depending on the size of the larger particles and the testing requirements. These are normally retained in large polythene bags.

Even though the samples are disturbed, the small-scale structure of clays, including features such as fissures, thin sand layers or varves (thin partings of silt or fine sand found in glacial clays) will still often be evident, making them useful to aid description and assessment of properties.

B.2.2   Open-Drive Samplers

Open-drive samplers are usually used in boreholes but may also be used in trial pits, as discussed earlier in Section B.1.1. The most common type is the 100 mm (nominal) diameter sampler but smaller samplers, typically 38 mm or 40 mm, are used for hand sampling, usually in trial pits or auger

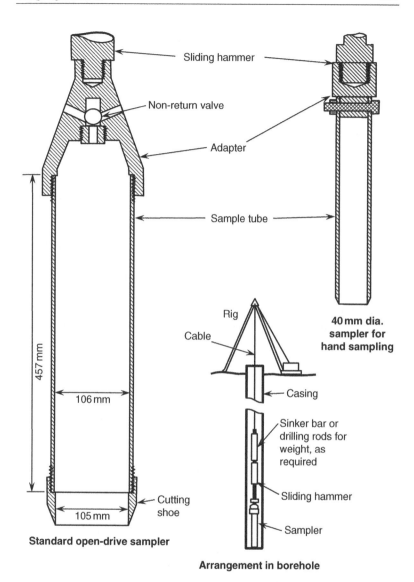

Figure B3   Schematics of open drive samplers.

holes. The essential features of the samplers and ancillary equipment are shown in Figure B3.

A cutting shoe protects the end of the 100 mm sampler and an adapter protects the top and connects it to a sliding hammer that is repeatedly raised and lowered to drive it into the ground. One or two 'sinker bars' or drilling rods may be added above the hammer to add weight and aid driving in stiff to hard soils. In a light cable percussion boring, raising and releasing the sliding hammer is controlled by the driller, as is the distance the sample tube is pushed into the ground, so the operation relies on his skill not to raise the hammer too high, pulling back the sample tube, or to overdrive it, compacting the sample. There is a temptation for drillers to overdrive the tube to ensure they get a full sample, especially as partial samples are often not paid for at the full rate, if at all.

Another problem that can arise is that, in stony ground, cutting shoes become damaged. Since they are expensive to replace and many drillers work as subcontractors, owning and maintaining their own equipment, there is an understandable temptation to keep on using even badly distorted cutting shoes. This must not be allowed as it will reduce sample quality.

The number of blows required to drive the tube is recorded. This can give a very rough indication of the consistency of the ground, but it must be remembered that the effective hammer weight will depend on whether or not sinker bars have been used.

The risk of the sample falling out of the tube is said to be reduced by leaving the tube in the ground for a few minutes to allow the sample to swell before withdrawing it, since the cutting shoe is slightly narrower than the sample tube to aid sample retrieval. Some drillers do this but others pull out the tube as quickly as possible, with seemingly similar success in retrieving samples. A method used in soft clays and clayey sands is to connect two or three sample tubes together, drive them to the full depth, then take the sample from the upper or centre tube.

Variations of the simple open-tube sampler, developed to help retain the sample on withdrawal, include spring grab devices at the lower end that fold back as the sample enters and are pushed out as the sample tries to fall back down the tube, holding it in place.

As described previously, sample tubes may be used in trial pits, driving them with a sledge hammer or using the bucket of the excavator.

On recovery, the sample tube ends are cleaned out and the soil protected from drying out by pouring melted paraffin wax over it.

In the laboratory, samples are extruded, as required for testing, by pushing them out of the top of the tube. If samples of stratified soil are fully extruded and cut cleanly down their length, then significant curving of the strata will be evident, caused by friction between the soil and tube; an indication of sample disturbance, even in this 'undisturbed' sample. Disturbance can be reduced by using tubes that take separate liners, usually uPVC.

B.2.3   Piston Samplers

Piston samplers were developed to overcome the problems of sample retention and sample disturbance described above for open-drive samplers in soft or sandy deposits. The principal features of the piston sampler are illustrated in Figure B4a.

The sampler and piston are pushed through any disturbed ground, with the piston at the bottom of the sample tube as shown in Figure B4b. The sample tube is then pushed into the ground, with the piston maintained at constant depth as shown in Figure B4c. The piston and tube are then withdrawn together. Contact of the sample with the piston, with no air present, helps keep it in the tube. Piston samplers are usually jacked into the ground rather than driven and are typically made of smooth-finished stainless steel rather than mild steel as with most open-drive samplers, both of which help reduce sample disturbance.

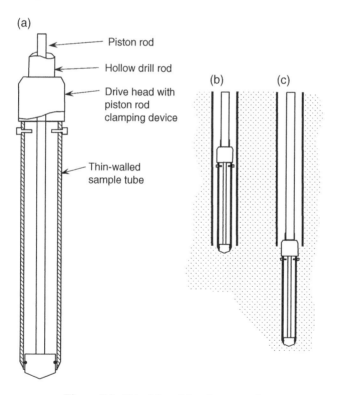

(a)

Piston rod

Hollow drill rod

Drive head with piston rod clamping device

Thin-walled sample tube

(b)

(c)

Figure B4   Principles of the piston sampler.

B.2.4   The Standard Penetration Test

The standard penetration test (SPT) is used to determine the relative density of granular soils and to infer other properties, including shear strength and compressibility.

Before starting the test, the bottom of the hole must be cleaned out to remove any disturbed material. Boring techniques that can adversely affect test results are discussed above in Section B.1.2.

A standard split spoon sampler, shown in Figure B5, is driven 450 mm into the soil at the base of the borehole by repeated blows of a hammer (known as a 'monkey') of standard weight and drop distance. The arrangement of the sampler and driving mechanism is shown in Figure B6. The blows are recorded every 75 mm but, in calculating the N-value, the blows for the first 150 mm (the 'seating blows') are ignored and the N-value reported as the total of the blows for the remaining 300 mm. The only exception to this is where penetration does not significantly exceed 150 mm even after 50 blows, when the N-value is reported as 'including seating blows'. For any penetration below 450 mm, the penetration achieved should also be given alongside the N-value.

In sands, the open driving shoe should be used, so that a small disturbed sample is obtained from it within the split spoon sampler, but in gravels the material is too coarse to enter the narrow sampling tube, and the shoe would be damaged, so it is exchanged for a cone tip, as illustrated, with no sample obtained. The N-value is usually assumed to be equivalent for either the split spoon sampler or the cone. Standard penetration testing is also carried out in

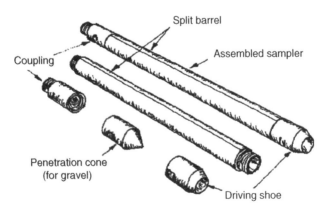

Figure B5   Split spoon sampler and cone attachment.

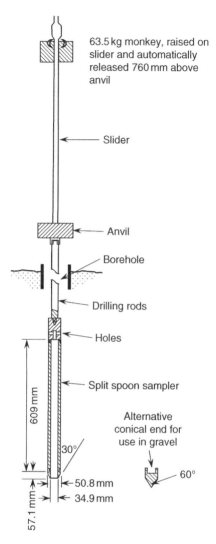

Figure B6 Schematic arrangement of standard penetration test equipment in borehole.

overconsolidated glacial clays which are difficult to sample because of their very stiff or hard consistency and the presence of gravel and cobbles.

Because of the risk of damage to the open driving shoe then, as with the cutting shoe of the open-drive sampler described above, there is a temptation for drillers to continue using damaged equipment, or to always use the cone tip so that no samples are obtained regardless of the soil type. This should not be permitted.

B.3 Probes

Probes dispense with the need to form a borehole or retrieve samples by assessing the soil properties *in situ*. Essentially probes come in two types:

1. *dynamic probes*, which work on a similar principle to the standard penetration test, using a standard falling hammer to drive a rod with a cone-shaped tip through the soil; and
2. *static probes*, in which a cone-tipped rod is pushed through the soil at a constant rate, normally with load cells recording tip and side resistance, and possibly water pressure.

B.3.1 Dynamic Probes

These are the simplest form of probe and involve no complex equipment, resistance being measured simply as the number of blows to attain a given penetration, usually every 100 mm. Cone dimensions and weight and drop distance of the hammer weight vary, but they can be correlated with SPT *N*-values to obtain shear strength and relative density as described in Chapters 3 and 6. The most popular probe is now the super-heavy probe which has the same hammer weight and drop distance and the same cone dimensions as the standard penetration test, so the number of blows per 300 mm can be used directly as an SPT *N*-value without the need to make assumptions about correlation factors.

The probe is progressed through the ground by adding further driving rods, usually 1 m in length. The driving rods are normally slightly thinner than the cone tip, to reduce friction on the rod sides. However, skin friction on the rods can affect results, compared with the standard penetration test, effects becoming more marked with depth. Procedures usually require the rods to be rotated by at least one complete turn each time a new rod is added, to help reduce skin friction. Clearly, there is some reliance here on the drillers carrying out good practice.

Dynamic probes can penetrate most clays and loose and medium dense sands to depths of up to about 9 m, though 6 m is a more usual maximum depth, depending on equipment and ground conditions. Progress is difficult through very stiff or gravelly clays, and the probe will generally be stopped by dense gravels, very stiff and hard clays and obstructions.

Since no soil samples are recovered, the soil type must be inferred from the blow counts. Because of this lack of soil definition, dynamic probes are often used in conjunction with traditional boring methods to check ground variation between boreholes and aid in interpretation of the results.

B.3.2   Static Probes

With static probes, a cone is pushed into the soil at a constant rate. The conical tip and rod sides (the 'sleeve') are separate so that cone resistance and tip resistance can be measured separately. With early mechanical cones the cone tip was pushed in first and cone resistance measured, then the sleeve was pushed in to measure sleeve resistance. With modern electronic cones the tip and sleeve are pushed in together at a constant rate of penetration, tip and sleeve resistance being measured simultaneously. As with dynamic probes, static probes come in a variety of sizes, but the tendency for some time has been to standardise on the Dutch probe of the type shown in Figure B7 (details in Table B1). The cone requires electronic equipment to record results, and both cone and supporting equipment are normally contained in a heavy truck which also acts as a kentledge (counterweight) during penetration. The trucks are often equipped with ground anchors to increase their effective weight.

The Dutch cone was developed for the poorly consolidated clays, silts and sands that predominate in Holland, and is eminently suitable for these types of soil, but it cannot be used in stiffer clays or dense sands and gravels and cannot penetrate obstructions. However, probes can be used through bands of coarse or dense material by pre-drilling; pushing a dummy rod 45-50 mm diameter through the layer ahead of the cone to reduce resistance, but this means that the properties of the coarse band are not measured. The penetration achievable depends on the cone capacity and soil type, but typically cones can be expected to penetrate up to about 15 m depth.

B.3.2.1   Principal Cone Types

There are two principal types of operation of cones:

1. *subtraction cones* measure the sleeve resistance and the combined sleeve and tip resistance, the tip resistance then being calculated as the difference between the two values; while
2. *compression cones* measure the tip and sleeve resistance separately.

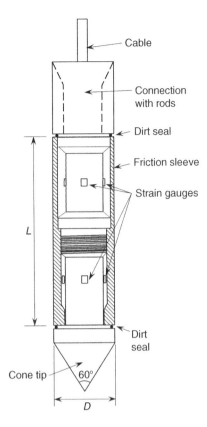

Figure B7 Principal features of a static cone.

Table B1 Standard cone sizes and attributes.

	Cone capacity	
	10 tonne	15 tonne
Diameter D (mm)	35.7	43.8
Projected end area of cone (mm^2)	1000	1500
Length of friction sleeve L (mm)	134	164
Area of friction sleeve (mm^2)	15000	22500
Max force on penetrometer (kN)	100	150
Max cone resistance q_c (if $f_s=0$) (MPa (kN))		
– subtraction cone	100 (100)	100 (150)
– compression cone	100 (100)	100 (150)
Max sleeve friction f_s (if $q_c=0$) (MPa (kN))		
– subtraction cone	6.6 (100)	6.6 (150)
– compression cone	1.0 (15)	1.0 (22.5)
Diameter of push rods (mm)	36	36
Rate of penetration (mm/s)	20	20
Maximum inclination (°)	15	15

With compression cones, there is an upper limit (typically about 1 Mpa) on the sleeve resistance, whereas there is no restriction on sleeve resistance with a subtraction cone. It is therefore generally recommended that subtraction cones be used where stiff clays are likely to be encountered, since these tend to generate high sleeve resistance.

Table B1 gives the sizes of the two most common capacity cones.

B.3.2.2   Piezocones

Piezocones have the additional facility of measuring pore-water pressure. This can be measured at various locations on the cone tip or the sleeve, depending on cone design, but it is usually recommended that it be measured just behind the cone tip. Measuring pore-water pressure gives more reliable results, both in identifying soil type and in evaluating shear strength and deformation properties.

The progress of the cone through the soil causes a build-up of pore-water pressure. In the dissipation test, the cone is halted and the pore-water pressure is recorded over time, as it dissipates. Time is normally plotted to a logarithmic or square-root scale, as for the consolidation test. The value of the coefficient of consolidation c_h (h for horizontal drainage) can be obtained from t_{50}, the time for 50% consolidation as described in Chapter 5.

It is essential for accurate measurement that the pore pressure sensor, which is a stiff membrane-type transducer, be fully saturated throughout the test. This is normally done by immersing it in de-aired, distilled water. In unsaturated soils and dilative soils such as dense sands, glycerine or silicone oil are used to help maintain saturation. The filter element is typically covered in a thin rubber membrane to maintain saturation of the tip during insertion; this is broken when the penetrometer moves through the soil. Various methods are used to achieve and maintain saturation, the tip often being prepared in a saturated condition in the laboratory beforehand.

B.3.2.3   Soil Sampling

It has been noted previously that a disadvantage of probing over conventional drilling is that no samples are obtained for inspection to confirm the soil type. This limitation can be overcome to some degree by use of a MOSTAP (Monster Steek Apparaat) sampler, which consists of a sample tube with a stocking liner and a separate cone at the bottom. The cone and tube are pushed to the required sampling depth, then the tube pushed further into the soil while the cone remains static, travelling up inside the tube. Its operation is therefore analogous to that of a piston sampler, described previously. Two sample diameters are used: 35 mm, yielding significantly disturbed samples; and 65 mm, yielding what may be regarded as undisturbed samples for testing purposes.

B.3.2.4   Interpretation of Results

Various charts exist to identify soil type based on cone resistance and sleeve resistance. Figure B8, published by Lunne *et al.* (1997), shows a plot of soil types correlated against a combination of cone resistance and 'friction ratio'; the ratio of cone resistance to sleeve resistance, expressed as stresses (MPa).

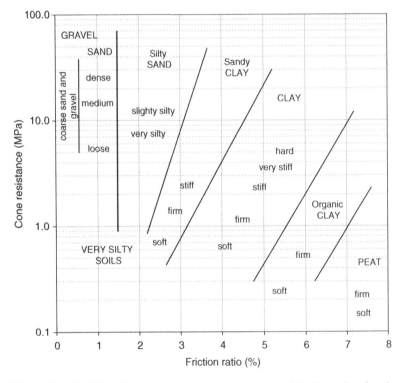

Figure B8   Soil identification based on cone resistance and friction ratio of static cone tests. After Lunne *et al.* (1997).

Cone tip resistance and friction ratio values can also be used in conjunction with charts to estimate SPT N-values and relative density, consolidation properties and shear strength parameters, as discussed in Chapters 3, 5 and 6, respectively. However, these rely on good correlations between resistance values and soil properties. Some design methods have been developed that use cone resistance values directly but, again, they ultimately rely on a good correlation between resistance values and soil properties.

Reference

Lunne, T., Robertson, P. K. and Powell, J. J. M. 1997. *Cone Penetration Testing in Geotechnical Practice*. Spon Press, London.

Index

Soil Properties and their Correlations, Second Edition. Michael Carter and Stephen P. Bentley.
© 2016 John Wiley & Sons, Ltd. Published 2016 by John Wiley & Sons, Ltd.

Printed and bound by CPI Group (UK) Ltd, Croydon, CR0 4YY

27/10/2024

14580349-0001